# Claims on highway contracts

under the *5th Edition of the ICE Conditions of Contract* and the *7th Edition of the Manual of Contract Documents for Highway Works*

Dr Robert N. Hunter

Published by Thomas Telford Publishing, Thomas Telford Services Ltd, 1 Heron Quay, London E14 4JD

First published 1997

Distributors for Thomas Telford books are
*USA*: American Society of Civil Engineers, Publications Sales Department, 345 East 47th Street, New York, NY 10017-2398
*Japan*: Maruzen Co. Ltd, Book Department, 3–10 Nihonbashi 2-chome, Chuo-ku, Tokyo 103
*Australia*: DA Books and Journals, 648 Whitehorse Road, Mitcham 3132, Victoria

The cover photograph was supplied courtesy of West Lothian Council and the Forth Local Authority Consortium.

A catalogue record for this book is available from the British Library

**Classification**
*Availability*: Unrestricted
*Content*: Original analysis
*Status*: Author's invited opinion
*User*: Engineers, academics and students

ISBN: 07277 2580 7

© Dr Robert N. Hunter, 1997

All rights, including translation reserved. Except for fair copying, no part of this publication may be reproduced, stored in a retrieval system or transmitted in any form or by any means, electronic, mechanical, photocopying or otherwise, without the prior written permission of the Books Publisher, Thomas Telford Publishing, Thomas Telford Services Ltd, 1 Heron Quay, London E14 4JD.

This book is published on the understanding that the author is solely responsible for the statements made and opinions expressed in it and that its publication does not necessarily imply that such statements and/or opinions are or reflect the views or opinions of the publishers.

Every effort has been made to ensure that the statements made and the opinions expressed in this publication provide a safe and accurate guide; however, no liability or responsibility of any kind can be accepted in this respect by the publishers or the author.

Typeset in Great Britain by Keytec Typesetting Ltd, Bridport, Dorset.
Printed and bound in Great Britain by Redwood Books, Trowbridge, Wiltshire.

# Preface

This book has been written by a civil engineer who finds himself involved in claims from time to time. There are many books on claims but few, if any, which have been written by civil engineers. As explained in the text, if the matter is not settled by the time that arbitration becomes a real possibility then it is time to consult a lawyer skilled in contract affairs. Up to that time, civil engineers are usually on their own. This book has been written to provide guidance in such circumstances.

On a personal level, I do wish that more civil engineers would regard claims for what they truly are: a fundamental necessity. I enjoy claims. The intellectual sparring is stimulating, particularly if one's opponent is skilled and experienced and does not engage in fatuous counter-arguments. Too many people treat claims as a personal insult. Life is too short so to do.

*Robert N. Hunter*

# Acknowledgements

Very many thanks to Joan Morrell and Jacqueline Watt who typed in the vast bulk of the Clauses from the *5th Edition of the ICE Conditions of Contract*; Ian Walsh of Kent County Council for lending me early versions of the *Manual of Contract Documents for Highway Works* (I did write to the Highways Agency but received no reply); Mike Pullen of Staffordshire Engineering Consultants for his enormous assistance with the publication history of the *Manual of Contract Documents for Highway Works*; Jack Edgar for reading the final draft and providing stimulating discussion on my opinions; Julian Cruft for checking part of the final draft; West Lothian Council for giving me access to a substantial proportion of the many texts used in preparing a book of this nature; Acorn Computers and Computer Concepts for producing hardware and software which makes word processing an absolute joy; and everyone else who assisted me in this project.

# Contents

| | | |
|---|---|---|
| **1.** | **Introduction** | **1** |
| 1.1. | Preamble | 1 |
| 1.2. | The philosophy of claims | 1 |
| 1.3. | Avoiding claims | 4 |
| 1.4. | Types of claim | 6 |
| | 1.4.1. *Quantum meruit* | 6 |
| | 1.4.2. Breach of a common law duty in tort | 6 |
| | 1.4.3. *Ex gratia* | 7 |
| | 1.4.4. Common law breach of contract | 7 |
| | 1.4.5. Contractual | 8 |
| 1.5. | The nature of claims | 8 |
| 1.6. | The *5th Edition of the ICE Conditions of Contract* and the *7th Edition of the Manual of Contract Documents for Highway Works* | 10 |
| 1.7. | References | 13 |
| **2.** | **The *5th Edition of the ICE Conditions of Contract*: Clauses which may affect claims** | **14** |
| 2.1. | Preamble | 14 |
| 2.2. | Introduction | 16 |
| 2.3. | Clause 5 | 17 |
| 2.4. | Clause 7 | 19 |
| 2.5. | Clause 11 | 23 |
| 2.6. | Clause 12 | 25 |
| 2.7. | Clause 13 | 33 |
| 2.8. | Clause 14 | 37 |
| 2.9. | Clause 17 | 43 |
| 2.10. | Clause 20 | 45 |
| 2.11. | Clause 27 | 48 |

| | | |
|---|---|---|
| 2.12. | Clause 31 | 53 |
| 2.13. | Clause 36 | 55 |
| 2.14. | Clause 38 | 58 |
| 2.15. | Clause 40 | 60 |
| 2.16. | Clause 41 | 63 |
| 2.17. | Clause 42 | 63 |
| 2.18. | Clause 43 | 66 |
| 2.19. | Clause 44 | 67 |
| 2.20. | Clause 47 | 71 |
| 2.21. | Clause 48 | 76 |
| 2.22. | Clause 50 | 81 |
| 2.23. | Clause 51 | 82 |
| 2.24. | Clause 52 | 86 |
| 2.25. | Clause 55 | 96 |
| 2.26. | Clause 56 | 98 |
| 2.27. | Clause 59A | 101 |
| 2.28. | Clause 59B | 107 |
| 2.29. | Clause 60 | 112 |
| 2.30. | Clause 61 | 120 |
| 2.31. | Clause 66 | 122 |
| 2.32. | Summary of provisions of Clauses | 127 |
| 2.33. | Summary of claims procedure | 132 |
| | 2.33.1. Introduction | 132 |
| | 2.33.2. What are costs? | 132 |
| | 2.33.3. Claims for extensions of time | 133 |
| 2.34. | Recent legislation to provide for adjudication | 133 |
| 2.35. | Amendments to the *ICE Conditions of Contract* | 138 |
| 2.36. | References | 139 |

**3. The *7th Edition of the Manual of Contract Documents for Highway Works*: its constituents and functions** — **140**

| | | |
|---|---|---|
| 3.1. | Preamble | 140 |
| 3.2. | Introduction | 141 |
| 3.3. | Volume 0: 'Model Contract Document for Major Works and Implementation Requirements' | 143 |
| | 3.3.1. Section 0: Introduction of Manual System | 144 |
| | 3.3.2. Volume 0, Section 1: Model Contract Document for Highway Works | 145 |
| | 3.3.3. Volume 0, Section 2: Implementing Standards | 155 |
| 3.4. | Volume 1: 'Specification for Highway Works' | 157 |
| | 3.4.1. General | 157 |
| | 3.4.2. Series 000 – Introduction | 158 |

|  |  | 3.4.3. Series 100 – Preliminaries | 160 |
| --- | --- | --- | --- |
| 3.5. | Volume 2: 'Notes for Guidance on the Specification for Highway Works' | | 160 |
|  | 3.5.1. | General | 160 |
|  | 3.5.2. | Series NG 000 – Introduction | 161 |
|  | 3.5.3. | Series NG 100 – Preliminaries | 166 |
| 3.6. | Volume 4: 'Bills of Quantities for Highway Works' | | 167 |
|  | 3.6.1. | General | 167 |
|  | 3.6.2. | Section 1 – Method of Measurement for Highway Works | 169 |
|  | 3.6.3. | Section 2 – Notes for Guidance on the Method of Measurement for Highway Works | 172 |
|  | 3.6.4. | Section 3 – Library of Standard Item Descriptions for Highway Works | 174 |
| 3.7. | Claims associated with the *Manual of Contract Documents for Highway Works* | | 175 |
|  | 3.7.1 | General | 175 |
|  | 3.7.2. | Claims emanating from the Specification | 177 |
|  | 3.7.3. | Claims emanating from the Bill of Quantities | 177 |
|  | 3.7.4. | Claims emanating from the Method of Measurement | 181 |
| 3.8. | References | | 183 |

## 4. The claims process  184

| 4.1. | Preamble | 184 |
| --- | --- | --- |
| 4.2. | Foreplay | 184 |
| 4.3. | Presentation of claims | 187 |

## 5. Model claim  188

| 1. | Preamble | 190 |
| --- | --- | --- |
| 2. | Contract details | 191 |
| 3. | Nature of claim: claim related to adverse physical conditions in the form of unacceptable material below the foundation of a box culvert | 191 |
| 4. | Evaluation of claim | 194 |
| Appendix 1. | Sketch of the layout of the Works | 201 |
| Appendix 2. | Original Clause 14(1) programme | 202 |
| Appendix 3. | List of drawings | 203 |
| Appendix 4. | Correspondence related to claim | 204 |
| Appendix 5. | Programme showing actual progress | 216 |
| Appendix 6. | Dayworks sheets | 217 |

| | |
|---|---|
| Appendix 7. Non-productive overtime costs | 220 |
| Index of Standard Letters | 221 |
| General Index | 224 |

# 1
## Introduction

### 1.1. Preamble

Most civil engineering projects carried out in the United Kingdom which are concerned with the construction or maintenance of highways employ the *5th Edition of the ICE Conditions of Contract* and the *Manual of Contract Documents for Highway Works*. The *5th Edition of the Conditions of Contract* is a 'standard form' of contract first published in 1973 and revised in January 1979. Although it was ostensibly superseded by the publication of the 6th Edition in 1991, it continues to enjoy widespread use in civil engineering generally. The *Manual of Contract Documents for Highway Works* is now in its 7th Edition and is a coordinated suite of documents which control the specification of highway works and the means by which such works are measured and, therefore, paid.

In order to maximise returns in construction works (or often, nowadays, to minimise losses) it is necessary to have a full understanding of these documents, particularly those elements which have a financial connotation. The purpose of this book is explore those parts of the *5th Edition of the ICE Conditions of Contract* and the *Manual of Contract Documents for Highway Works* which are particularly important in the commercial aspects of highway contracts and to examine how they apply in claims situations. Those who understand the documents in detail will be best placed to derive the maximum benefit from the myriad of complex situations which occur as a result of such works being undertaken.

### 1.2. The philosophy of claims

Chambers defines claims as 'v.t. to call for: to demand as a right: to maintain or assert. -n. a demand for something supposed due: right or ground for demanding'. In civil engineering, claims are a fact of life. Most contracts are let on the basis of admeasurement i.e. the amount paid is calculated on the basis of the amount of work done. The exact quantities involved in Works of civil engineering are notoriously difficult to predict even when substantial site investigation work is carried out. This is undoubtedly due to the fact that much of the Works involves operations below ground and is subject to the vagaries of weather conditions. Thus, there has to be some apparatus for ensuring that the Contractor is paid equitably for executing the Works and claims constitute just such a mechanism. The philosophy behind both the *5th Edition of the ICE Conditions of Contract* and the *Manual of Contract Documents for Highway Works* is that the Contractor can, with confidence, price for the situations described in the contract in the full knowledge that there exists, within the contract documentation, remedies to ensure fair payment. Thus the promoter of the Works, or Employer as he is more normally described, derives the benefit of the minimum price at the outset.

A major feature of civil engineering is excavating in ground where the characteristics are largely unknown. Site investigations can assist but they are hindered by the fact that it is only feasible in such analyses to examine a small proportion of the soil conditions. The science of statistics requires only a relatively small sampling regime in order to predict, with some confidence, the nature of the entire population—look at the polls predicting the outcome of the 1992 United Kingdom General Election. However, a small sampling regime is only valid for the whole if it is consistent and frequently that is not the case—look at the polls predicting the outcome of the 1992 United Kingdom General Election. Soils are notoriously inconsistent in terms of both the type of soil and its condition i.e. moisture content and other physical characteristics. Another major factor in civil engineering is the prevailing weather which is equally unpredictable.

There are a number of other factors which may well produce contract documents which do not precisely specify the nature of the Works. The following list summarises all the major factors which may confound the preplanning of modern civil engineering projects

(*a*) soil characteristics
(*b*) prevailing weather conditions
(*c*) the size and complexity of modern civil engineering projects
(*d*) the interaction of different disciplines—highway engineers, landscape architects, lighting engineers, structural engineers, etc.

(*e*) elements of the Works are largely outwith the control of the Engineer or the Employer, e.g. utilities
(*f*) commercial or other pressures leading to inadequate drafting time
(*g*) commercial or other pressures leading to specification and/or method of measurement drafting being carried out by less senior and/or less skilled (i.e. less expensive) staff.

The Employer, quite understandably, will wish to know, within fairly close limits, what a construction project is likely to cost. However, when the precise nature is not known, the Contractor cannot provide an accurate quotation, and a financial regulator will be necessary to evaluate the unforseen or unknown elements in the project. That regulator is claims. Without that regulator, Contractors will simply make a guess about the degree of unknown elements in the Works which will, as often as not, lead either to bankruptcy or vast profits and neither situation is desirable, in the longer term, either to the Contractor or those who rely upon the contracting fraternity to construct their projects. Despite the fact that many people are unhappy about the way that civil engineering contracts operate, the truth is that there really is no other feasible option.

The fact that most standard forms of contract, ICE or JCT *Conditions of Contract* for example, recognise and set down procedures for pursuing claims is testament to the propriety of the concept and practice of paying for extras via claims. Sadly, many Engineers and other professionals associated with civil engineering still regard claims as a great iniquity. Hence, the plethora of non-standard forms, a tendency which carries substantial dangers because of the possible financial consequences. It is difficult to be sympathetic to those who pay substantial sums in claims because the contract has been poorly drafted. It is not difficult to minimise the possibility of claims. In highways contracts, one obvious approach is to prepare amendments to the specification and method of measurement in strict accordance with the rules set down in those documents. In the case of the documents featured in this book, the rules, which are set out in detail in the documents themselves, are simple to apply and yet they are frequently ignored. The Employer blames a claims orientated Contractor when in fact the reality is that it was the documents which gave the Contractor misleading information and it was this spurious data which played a major part in influencing the pricing strategy. Hence when this information turns out to be wrong then so too does the pricing regime and the Contractor has no option but to attempt to recover outlay.

Claims arise from one of two fundamental reasons. Indeed, from what has been stated above it is possible to conclude that they are

(a) because the nature of the contract changes due to certain parameters which were not as envisaged or
(b) because of errors in drafting.

Errors in drafting include situations where the interpretation of contract-specific requirements is capable of more than a single interpretation. Each requirement in the contract has to be unambiguous if claims are to be avoided. Some sympathy has to be accorded to those involved in drafting since it can be very difficult to ensure that the only plausible interpretation is that which was intended by those who compiled the documents.

## 1.3. Avoiding claims

Those responsible for drafting contracts can play a substantial role in producing documents which minimise the possibility of claims. There is absolutely nothing wrong with such an approach. As has been seen above, there are basically only two reasons why claims arise: misrepresentation in the documents and changes in circumstances. If those drafting contracts bear in mind these two principles they can do much to avoid claims.

In modern civil engineering contracts, it is often the case that certain requirements have a number of different interpretations. It is appreciated that producing a form of words which are capable of only one interpretation is difficult. For example, the specification on a dual carriageway refurbishment contract stated that the Contractor must maintain two lanes open for traffic at all times during the execution of the Works. What the specifier meant was that two lanes of the eastbound carriageway and two lanes of the westbound carriageway must be open at all times during the execution of the Works. The Contractor priced traffic safety and control on the basis of having one eastbound lane and one westbound lane open for traffic at all times i.e. two lanes open for traffic on the dual carriageway at all times during the execution of the Works. Did the Contractor see this ambiguity at the time of tender? The answer is academic; the interpretation is perfectly reasonable in the circumstances but the Contractor should never have had the opportunity to attribute such a meaning thereto. The Clause should have been framed unambiguously. Sometimes this is at the cost of brevity but that is unimportant if it avoids a claim. It is conceivable that the Contractor did see this ambiguity during the tendering stage but it is equally possible that the other interpretation was not clear. If the Contractor did, should clarification have been sought and if so, why? He may have felt that the reading gave a commercial advantage by, very reasonably, pricing it on the basis of this interpretation when one or more competitors may have priced it on what would probably be a more expensive alternative viewpoint. Furthermore, at the time of tender, there is no contractual relationship between the parties and consequently no compulsion to seek clarification when there are ambiguities. Besides, where ambiguities exist, the courts may well give the Contractor the benefit of his interpretation—see the discussion of the doctrine of contra proferentem[1] in Chapter 2. There is a further factor which the Contractor is likely to have taken into account at the time of tender when encountering a poorly drafted contract. He usually wants to win the work and does not want to create an impression in the mind of the Engineer of being difficult for fear of forfeiting the contract. Those drafting documents should ensure that their requirements are unambiguous.

In quantitative terms, persons drafting contracts should try, as far as possible, to produce individual items which contain accurate quantities. They should also take great care in dealing with non-standard items. Non-standard items (or 'rogue items' as they are sometimes called) are those which are not found in the standard Method of Measurement for Highway Works and thus have to be drafted especially for a particular contract. They are often ill-defined and frequently lead to disputes (and sometimes but not always to claims) during execution of the Works. They should be prepared to conform rigorously with the rules used to produce the standard items. As stated earlier, those rules are set out in the *Manual of Contract Documents for Highway Works* and are simple to apply.

It is suggested that the Contractor who is pricing a tender where there are major omissions in the item coverages, inconsistent errors in formulating preambles to non-standard items and numerous areas of ambiguity find the situation both frustrating and difficult. Contrary to popular belief, what he never does is rub his hands with glee at the prospect of a large number of claims. The outcome of claims is never certain and he will want to win more work in future. Most Contractors find it infinitely preferable if the contract is properly framed and the intentions of the Engineer are clear and unambiguous and the quantities are the result of considered intelligent production by skilled and experienced staff. Any claims which then arise are the result of conditions that could not reasonably have been foreseen (what a useful phrase) by the Engineer at the time of design and thus, since resultant claims are unlikely to be seen as inherent criticism of his design or contract preparation, he is more likely to be truly objective in their assessment. Contractors do not, by and large, relish contracts which are framed around a set of documents and quantities which are poorly drafted.

## 1.4. Types of claim

On a legal basis, it is convenient to divide claims into a number of different categories. These are

(a) *quantum meruit*
(b) breach of a common law duty in tort
(c) *ex gratia*
(d) common law breach of contract
(e) contractual.

### 1.4.1. Quantum meruit

The phrase *quantum meruit* means 'what it is worth'. It means that payment is on a fair and reasonable basis. In civil engineering, contracts paid on such a basis are extremely rare although payment may turn out to be made on such a basis. This book does not consider such claims in any detail but suffice it to say that if such a basis occurs then consideration of the content of later chapters should enable a fair valuation to be made.

### 1.4.2. Breach of a common law duty in tort

According to Hudson[2] tortious liability is a 'liability to pay damages which arises not out of contract, but from a wrongful act'. In Scotland, the word 'delict' would be used. Hudson cites as an example of the tort of negligence a situation where a motorist is sued for damages on the grounds that he drove negligently and, as an example of the tort of libel, the case where a newspaper is sued for defamatory statements.

There is a difference between the amount of damages in contract and tort. Under contract law, the damages are meant to reflect the amount of the benefit which the injured party would have received had the contract been performed. Under the law of tort, the damages are meant to restore the injured party to the position it would have been in had the wrongful act not occurred. Claims under tort would be relatively unusual in civil engineering contracts and are not covered in this book.

### 1.4.3. Ex gratia

The phrase *ex gratia* means 'out of gratitude'. The Contractor may seek payment for some event over which he had no control but for which there is no provision in the contract which would allow the Engineer to recognise the Contractor's difficulty and the situation does not involve a breach of any kind. An example would be where the price of oil, and hence bitumen, rose inordinately in a contract which had no price fluctuation clause or the price fluctuation provisions were such that they did not compensate for a significant rise in one material. In today's commercially orientated world, this type of claim is rarely paid.

### 1.4.4. Common law breach of contract

The law of contract is based on common law. In other words, there are few statutory provisions which deal with this area of commercial activity and the rules which govern the courts' view of specific events are often decided on the basis of previous rulings by the courts. These can be overturned by contemporary judgments but generally provide some guidance on the likely outcome in a particular situation.

Generally, a breach occurs when one of the parties fails to comply with some express or implied term of the contract. An express term is one which specifically details a right or obligation of one or both of the parties to a contract. An implied term is one that may be imported by an Act of Parliament, for example, the concept of fitness of purpose is an implied term on the basis of the Sale of Goods Act 1979 or it may be a term which is imposed by the courts in order to make the contract commercially effective. In civil engineering terms, there is an implied term that neither the Engineer nor the Employer will hinder the Contractor from carrying out the Works (*Merton London Borough v. Leach (Stanley Hugh)* (1985) 32 BLR 51). However, the courts will rarely agree that there is an implied term which contradicts an express term since it generally takes the view that specific terms which are clear and unambiguous reflect the basis upon which the parties did consciously contract and these terms were acceptable as applicable at the time when the contract was formed.

### 1.4.5. Contractual

This type of claim relates to breaches of specific requirements within the contract. They are found in many types of 'standard form'. Standard forms are those which occur commonly in civil engineering, ICE Conditions of Contract, ICE Minor Works Conditions etc.

These forms have been produced with standard provisions which recognise the need for extra payment to reflect certain fairly common events on construction sites. They provide a clear set of rules for the Contractor to follow if he wishes to claim extra time for the execution of the Works, generally referred to as an 'extension of time', or if he wishes to claim additional payment to cover increased work in executing the contract. Equally, they provide a framework for the Engineer to recognise construction difficulties not quantified or notified at the time of tender in the form of extensions of time and/or increased payment. These terms avoid the need for arbitration or litigation in the vast majority of cases. The great majority of construction contracts are settled amicably (or reasonably so) thus avoiding the need to resort to formal proceedings.

## 1.5. The nature of claims

Where there is a breach of contract the remedy is normally damages, i.e. a measure of the financial loss faced by the aggrieved party. Contractual damages are meant to place the party which has been wronged in the position which it would have enjoyed had the contract been executed in accordance with the terms of the contract. Assessment of damages is not an exact science which may explain why many Engineers find claims a difficult area of professional activity. After all, Engineers are used to calculating a stress in a structure very accurately; the calculation of claims which are contractual damages is a much less precise activity.

A Contractor makes a claim in order to recover losses whether actual or perceived. A claim is likely to seek to address two areas where the Contractor feels that he has an entitlement: a request for an extension of time and the recovery of costs and perhaps, depending on the circumstances, profit.

The reason for requesting an extension of time will be either to recover the costs associated therewith or to reduce the Contractor's liability for liquidated damages which have been withheld by the Employer or both. Liquidated damages are supposed to be a genuine pre-estimate of loss (*Clydebank Engineering and Shipbuilding Co v. Don Jose Ramos Yzquierdo-y-Castaneda* (1905) AC6).

The courts recognise the value of fixing a penalty for late completion at the time when the contract is made. This penalty, or liquidated damages as it is called, is a sum normally stated in the Appendix to the Form of Tender. It is a rate to be applied per day or per week of non-completion. It can be challenged by the Contractor on the basis that it is not a measure of damages, but a penalty, although the onus is on the Contractor to prove that it is punitive and not on the Employer to prove that it is a genuine pre-estimate of loss. The means of calculation is not fixed and the Employer can fulfil the requirement for it to be a genuine pre-estimate of loss in a number of ways. The fact that the Contractor does not challenge the figure at the time of tender is not evidence that it is acceptable to the Contractor and will not be seen by the courts as proof of its reasonableness. If it turns out that the Employer suffers no loss, this fact will not provide grounds for the liquidated damages to be set aside. The concept that the sum is not punitive does not mean that it is insignificant. It can often represent substantial sums of money, particularly where delay runs into months. Any Contractor who is commercially orientated will do everything possible to ensure that liquidated damages do not apply or that they are minimised. One of the standard elements of the type of contract used in civil engineering is the Appendix to the Form of Tender. It sets out some important facets of the contract and may contain, amongst other matters, the Time for Completion of the Works. Contractors should ensure at the time of tender that the period stated is a reasonable time frame for carrying out the Works. There is no precedent to suggest that the Contractor will be successful in trying to have the period for execution of the Works revised on the basis that this period, as stated in the tender, was too short. As suggested earlier, the courts will very rarely set aside what is included in the contract as an express term. They generally take the view that the parties were agreed on this term at the time of tender and, like most express terms, it forms an inviolable ingredient in the contract.

Most claims are settled long before arbitration or litigation become a reality. Those Contractors who operate in strict accordance with the rules, possess a complete understanding of the documents forming the contract and apply its provisions in a professional manner are most likely to impress the Engineer and achieve settlement at an early date. Hence, a fundamental understanding and knowledge of the standard contract elements enhances the likelihood of the Contractor achieving the desired result—a fair return for his investment.

## 1.6. The *5th Edition of the ICE Conditions of Contract* and the *7th Edition of the Manual of Contract Documents for Highway Works*

The purpose of this book is to examine both these documents in detail to see how they interact in forming a major proportion of highway contract documentation and, in particular, their roles in claims situations.

The history of the *ICE Conditions of Contract* is set out in Table 1.1.

An obvious question is why this book should examine the *5th Edition of the ICE Conditions of Contract* and not the *6th Edition*, a newer publication? The reason is that the *Manual of Contract Documents for Highway Works* is based on the use of the *5th Edition* and not the *6th Edition*. There is no obvious reason why this may be so although, as is about to be seen, the *7th Edition of the Manual of Contract Documents for Highway Works* was first introduced in December 1991 and perhaps its production timetable did not allow it to be amended to match the provisions of the *6th Edition* which was published some eleven months earlier. Certainly the *6th Edition* does not seem to be particularly different in legal terms from the *5th Edition*; it is written using language which is eminently more easily understood and that cannot be a negative consideration. Perhaps the

*Table 1.1. Publication history of the ICE Conditions of Contract*

| Edition | Date of Publication | Notes |
| --- | --- | --- |
| First Edition | December 1945 | Agreed by the Institution of Civil Engineers and the Federation of Civil Engineering Contractors |
| Second Edition | January 1950 | This and subsequent editions approved as above along with the Association of Consulting Engineers |
| Third Edition | March 1951 | |
| Fourth Edition | January 1955 | |
| Fifth Edition | June 1973 | Revised January 1979; reprinted with amendments January 1986 |
| Sixth Edition | January 1991 | |

*8th Edition of the Manual* which is due to be published around mid-1997 (as advised at the time of writing around mid-1996) will relate to the *6th Edition of the ICE Conditions of Contract.*

The *7th Edition of the Manual of Contract Documents for Highway Works* is the current version and its publication history is as follows. Technically, as can be seen from Table 1.2, these documents are not the 7th Edition, but as this is the term most commonly

*Table 1.2. (below and overleaf). The publication history of the Manual of Contract Documents for Highway Works*

**1st Version**
Consisted of

(a) 'Specification for Road and Bridge Works' (1951) (1st Edition)
(b) 'Notes on the Preparation of the Bill of Quantities' (1951).

Published by the Ministry of Transport and Civil Aviation. The associated Conditions of Contract were the Ministry of Transport *Standard Forms Conditions of Tender and Contract for Road and Bridge Works* published in 1947.

**2nd Version**
Consisted of

(a) 'Specification for Road and Bridge Works' (1957) (2nd Edition)
(b) 'Notes on the Second Edition of the Specification for Road and Bridge Works and on the Preparation of the Bill of Quantities' (1958).

Published by the Ministry of Transport and Civil Aviation. The associated Conditions of Contract were the *4th Edition of the ICE Conditions of Contract* published in 1955.

**3rd Version**
Consisted of

(a) 'Specification for Road and Bridge Works' (1963) (3rd Edition)
(b) 'Notes on the Third Edition of the Specification for Road and Bridge Works and on the Preparation of Bills of Quantities'.

Prepared by the Ministry of Transport and published by HMSO. The associated Conditions of Contract were the *4th Edition of the ICE Conditions of Contract* published in 1955.

Note that at this stage the Preparation of Bills of Quantities was generally in accordance with the *ICE Standard Method of Measurement of Civil Engineering Quantities* (1962).

*Table 1.2.—continued*

**4th Version**
Consisted of

(a) '(MoT) Specification for Road and Bridge Works' (1969) (4th Edition)
(b) '(DoE) Method of Measurement for Road and Bridge Works' (1971) (1st Edition).

The sources of documents may be confused at this time since Ministry of Transport (MoT) issues became Department of Transport (DoT) issues. There was an MoT (Highways Directorate) document (which was not published by HMSO) entitled *Notes for Guidance and Library of Standard Item Descriptions for the Preparation of Bills of Quantities for Road and Bridge Works* which referred to a *Ministry of Transport Method of Measurement*. It has not been possible to define the latter and it may well have been an internal document.

Prepared by the Ministry of Transport, the Scottish Development Department and the Welsh Office and published by HMSO. The associated Conditions of Contract were the *4th Edition of the ICE Conditions of Contract* published in 1955 as amended.

**5th Version**
Consisted of

(a) 'Specification for Road and Bridge Works' (1976) (5th Edition)
(b) 'Notes for Guidance on the Specification for Road and Bridge Works' (1976) (1st Edition)
(c) 'Method of Measurement for Road and Bridge Works' (1977) (2nd Edition)
(d) 'Notes for Guidance and Library of Standard Item Descriptions for the Preparation of Bill of Quantities for Road and Bridge Works' (1978) (1st Edition).

Each of the above was amended by individual documents, each termed Supplement No. 1, published in 1978.

All the above were prepared by the Department of Transport, the Scottish Development Department and the Welsh Office and published by HMSO. The ICE Conditions of Contract were specified but not the Edition. The *5th Edition of the ICE Conditions of Contract* published in 1973 is the document which would normally be used in conjunction with the above.

**6th Version**
Consisted of

(a) 'Specification for Highway Works' (1986) (6th Edition)
(b) 'Notes for Guidance on the Specification for Highway Works' (1986) (2nd Edition)

Table 1.2.—continued

**6th Version**—continued

(c) 'Method of Measurement for Highway Works' (1987) (3rd Edition)
(d) 'Library of Standard Item Descriptions for Highway Works' (1987) (3rd Edition)
(e) 'Highway Construction Details' (1987) (3rd Edition).

The Specification for Highway Works was published in eight parts (Parts 1 to 6, Part 7(i) and Part 7(ii)).

The Notes for Guidance on the Specification for Highway Works was published in six parts (Parts 1 to 6) with Amendment No. 1 being published in 1988.

An Addendum to the Method of Measurement for Highway Works was published in 1991.

All the above were prepared by the Department of Transport, the Scottish Development Department, the Welsh Office and the Department of the Environment for Northern Ireland and published by HMSO. The Conditions of Contract were specified as the *ICE Conditions of Contract*, and given the date of publication of the Specification, would be the *5th Edition* published in 1973 as amended.

---

employed to describe the current suite, the same convention is followed in this book.

This book devotes separate Chapters to an in-depth examination of those aspects of both the *5th Edition of the ICE Conditions of Contract* and the *7th Edition of the Manual of Contract Documents for Highway Works* which may come into play in relation to claims. It then examines the normal claims procedure and sets out an example of a detailed claims submission embracing several different heads of claim.

## 1.7. References

1. **Duncan Wallace I. N.** *Hudson's Building and Civil Engineering Contracts.* Sweet & Maxwell, London, 1995, 11th edn, **1.009**, **1.224**, **4.042**, **7.013**.
2. **Duncan Wallace I. N.** *Hudson's Building and Civil Engineering Contracts.* Sweet & Maxwell, London, 1995, 11th edn, **1.273**.

# 2

## The *5th Edition of the ICE Conditions of Contract*: Clauses which may affect claims

### 2.1. Preamble

Claims may relate to a number of Clauses but certain key Clauses regularly feature in claims and, in fact, Clause 52(4)(a) or (b) should *always* feature in any letter intimating a contractual claim. Claims come about for a wide variety of causes but these can conveniently be allocated into one of two categories.

The first category relates to situations where there has been a physical change from the circumstances which are described in the contract documents. It may be that in a contract involving earthworks the balance of acceptable to unacceptable material or the balance between different types of acceptable materials has changed (as it commonly does in both cases). This may have a financial consequence which impacts upon the approach which was used by the Contractor in pricing the tender for the Works. He may have to find a location for disposal or processing which is further from the operation's locus than that envisaged at the time of tender. Such a change may mean a greater proportion of material has to be imported and a higher unit rate may have to be paid either for purchase or for haulage. This, in turn, may affect the programme for the Works and thus have an impact upon the completion date. As a result, recovery during the original contract period may decline in proportion to the reduced outputs. During the period of the extension of the contract the Contractor may suffer diminished contribution to overheads.

The second category is one where some part of the Works was inadequately or incorrectly or ambiguously described in the contract documents. This may be in the specification or in the bill of

# THE 5TH EDITION OF THE ICE CONDITIONS OF CONTRACT | 15

quantities or, for some reason, related to the method of measurement/item coverage or indeed in some other part of the contract documentation. In Chapter 3, the relationship between the 'Specification for Highway Works', the 'Notes for Guidance on the Specification for Highway Works', the 'Method of Measurement for Highway Works' and the 'Notes for Guidance on the Method of Measurement for Highway Works' is explored. As discussed in Chapter 1, if the Engineer has not exercised great care in the preparation of the contract documentation then the Contractor may well have made assumptions which did not match the Engineer's intentions.

The reason why the categories described above often lead to successful claims is that either is likely to jeopardise the Contractor's fundamental approach to pricing individual items in his tender or, in some cases, the overall tendering strategy, rendering it wholly or partially invalid.

Claims are event driven and the Clause(s) invoked are a function of the nature of these events. What this chapter does is discuss the meaning of specific Clauses and their proper application in relation to events which occur during construction. In terms of the pursuit of claims, some of the Clauses are directly involved and others, while apparently only indirectly involved are in reality playing a substantial role in their prosecution. This chapter also provides examples of standard letters which illustrate how the conditions may be implemented by a Contractor. The Clauses are quoted for ease of reference. As has been seen, a Clause may consist of several Sub-Clauses. The entire Clause is quoted but particular Sub-Clauses are not discussed if they are not relevant to claims and exclusion as such does not impinge on the understanding of the Clause as a whole.

When invoking a particular Clause in practice, it is vital that the user checks the contract itself to see whether any of the Clauses to be invoked have been altered within other parts of the contract. Such changes are often accorded a separate headed section called 'Amendments to the Conditions of Contract' or similar. In the *Manual of Contract Documents for Highway Works*, amendments to the standard *5th Edition of the Conditions of Contract* are, generally, to be found in the 'Model Contract Documents' for England, Scotland, Wales and Northern Ireland as appropriate (these are Parts 2, 3, 4 and 5 respectively of Volume 0, Section 1). At the time of writing, the Scotland document is in use but not yet available via the normal publisher, HMSO. The version of Section 3 of Volume 0 'Model Contract Document for Major Works and Implementation Requirements' available at the time of writing (mid-1996) via HMSO has a cover date of August 1995 but (as is stated on the front) is a reprint of the November 1992 version. This document is no longer used for trunk roads works but may be so for other smaller contracts.

## 2.2. Introduction

In framing notices under any of the Clauses of the Conditions of Contract, it is important to adopt a common approach. The use of the same words or phrases from the particular Clause/s being invoked is sound practice because it avoids the charge that the Contractor is not attributing the correct interpretation to the Clause. It also obviates the possibility of misapplication to prevailing circumstances.

In issuing any letter it is also good practice to quote the Clause number/s being invoked. This habit advises the Engineer that the Contractor is aware of his rights (and presumably his obligations) under the contract conditions. Furthermore, the author has heard it said by an Engineer that he would disallow a claim if the Contractor did not quote the Clause number or used the wrong Clause number or misquoted the Clause, hardly an object lesson in objectivity. Finally, as one has to issue many notices under any contract (there are many notices which are required but which have no connection with claims) this practice helps educate the user (and colleagues and recipients).

Giving proper notice in a professional format becomes a habit. Once established, it is easy and avoids questions of repudiation on the basis of not complying with a 'condition precedent'. A condition precedent is an action which is necessary before a subsequent action can take place e.g. the requirement for the Contractor to give notice under Clause 12(1) that adverse physical conditions or artificial obstructions not reasonably foreseeable have been encountered is a condition precedent to the Engineer issuing appropriate instructions under Clause 12(2). The author always creates a database of all standard letters (not just those relating to claims) for each contract on computer. In most cases, it takes only seconds to add the details necessary to validate any letter. All the letters contained in this chapter (and in the claim in Chapter 5) adopt the philosophy discussed above.

This chapter examines the Clauses under which claims may be made or are commonly associated with claims. It examines the meaning of the Clause or Sub-Clause under consideration and gives examples of any notices dictated by their use. There is nothing sacrosanct about the wording of these letters as long as they achieve the specific aim e.g. give notice of the claim, prompt the Engineer to grant an extension etc. They must also comply with the conditions as and where appropriate.

The *5th Edition of the ICE Conditions of Contract* groups one or more Clauses under different headings, for example 'General Obligations' groups Clauses 8 to 35 inclusive, 'Workmanship and Materials' groups Clauses 36 to 40 inclusive and so on. Each individual Clause which has no Sub-Clauses has a single marginal heading (technically a marginal note). Clauses which consist of several Sub-Clauses have an individual marginal note applied to each Sub-Clause and in one case has an overall marginal note (Clause 27: Public Utilities Street Works Act 1950). Note that, according to Clause 1(3), the headings and marginal notes have no relevance to the meaning or interpretation of the *Conditions of Contract* or the contract.

## 2.3. Clause 5

### DOCUMENTS MUTUALLY EXPLANATORY

> The several documents forming the Contract are to be taken as mutually explanatory of one another and in case of ambiguities or discrepancies the same shall be explained and adjusted by the Engineer who shall thereupon issue to the Contractor appropriate instructions in writing which shall be regarded as instructions issued in accordance with Clause 13.

This Clause gives equal status to all the documents that form the contract. This is a very important point since it means that they should be interpreted as a collective whole. Some modern contracts give certain documents superior status, for example giving the *Conditions of Contract* precedence over all other contract documents. Unless the superior document is framed very carefully, this approach can lead to errors of description. It is, however, becoming common practice presumably to avoid claims based on ambiguity. However, proper document preparation of contract documents is a far more effective means of avoiding ambiguity.

The resolution of ambiguities or discrepancies referred to herein may be of a technical or a legal nature. Engineers should bear in mind that it is often wise to seek advice from a solicitor skilled in contract matters when dealing with the resolution of disputes which have a legal connotation since Engineers themselves usually do not have the benefit of extensive professional training in the area.

It is a fundamental principle of common law (as seen earlier, contract law is composed largely of common law judgments) that where parts of a contract are reasonably capable of several different interpretations then any of those interpretations is valid. It is also a feature of common law that where a contract requirement is ambiguous then the interpretation of the party which did not devise the contract wording will take precedence. Therefore, in the situations being considered here, it is invariably the Contractor's interpretation which will prevail. This is known as the doctrine of contra proferentem,[1] i.e. the interpretation of those who framed the contract terms carries less legal weight than the interpretation of those who provide the goods or services of which the contract is the subject. This is a very important concept. Different interpretations are very common and hardly surprising given the detail contained in any contract and the variety of meanings which may be conveyed by the same phrase in the English language. It is very often the case that the meaning which was meant to be communicated by the writer of the specification (or the method of measurement or indeed any other part of the contract documents) is very different from that which the Contractor attributes to the relevant sections of the contract documents. However, at least one case suggests that this doctrine may not apply where the Employer/Engineer uses a standard form which is largely unaltered.[2]

Many claims, often those of substantial value, are related to the ambiguity or alleged ambiguity of the interpretation of contract-specific requirements and consequently those framing contract wording need to exercise particular care during preparation. Equally, the Contractor should be particularly alert when examining contract-specific requirements either for the purposes of tendering or in order to execute the Works (site staff often do not see the documents until they are actually involved in undertaking the contract).

Some Contractors, lacking a contractual claim, write to the Engineer seeking clarification. This clarification may well then give the Contractor the access to a contractual claim via the provisions of Clause 13(3) rather than one under contract law. Giving the Contractor such an opening may not be avoidable but this may, sooner or later, be to the advantage of the Engineer and the Employer.

Claims which relate to this Clause require a notice issued under Clause 13(3) and an example of such a notice is given later, in the section relating to Clause 13(3).

## 2.4. Clause 7

### SUB-CLAUSE (1)—FURTHER DRAWINGS AND INSTRUCTIONS

The Engineer shall have full power and authority to supply and shall supply to the Contractor from time to time during the progress of the Works such modified or further drawings and instructions as shall in the Engineer's opinion be necessary for the purpose of the proper and adequate construction completion and maintenance of the Works and the Contractor shall carry out and be bound by the same.

### SUB-CLAUSE (2)—NOTICE BY CONTRACTOR

The Contractor shall give adequate notice in writing to the Engineer of any further drawing or specification that the Contractor may require for the execution of the Works or otherwise under the Contract.

### SUB-CLAUSE (3)—DELAY IN ISSUE

If by reason of any failure or inability of the Engineer to issue at a time reasonable in all the circumstances drawings or instructions requested by the Contractor and considered necessary by the Engineer in accordance with sub-clause (1) of this Clause the Contractor suffers delay or incurs cost then the Engineer shall take such delay into account in determining any extension of time to which the Contractor is entitled under Clause 44 and the Contractor shall subject to Clause 52(4) be paid in accordance with Clause 60 the amount of such cost as may be reasonable. If such drawings or instructions require any variation to any part of the Works the same shall be deemed to have been issued pursuant to Clause 51.

### SUB-CLAUSE (4)—ONE COPY OF DOCUMENTS TO BE KEPT ON SITE

One copy of the Drawings and Specification furnished to the Contractor as aforesaid shall be kept by the Contractor on the Site and the same shall at all reasonable times be available for inspection and use by the Engineer and the Engineer's Representative and by any other person authorised by the Engineer in writing.

Sub-Clause (1) empowers the Engineer to issue further information in the form of drawings or instructions and requires the Contractor to obey these instructions.

Sub-Clause (2) requires the Contractor to give sufficient forewarning that information is required for the proper execution of the contract. Any such issue is deemed to have been issued under Clause 51 which relates to ordered variations and means that it constitutes, of itself, an ordered variation issued in the form of the drawing or specification. It may indeed constitute several variations and thus several ordered variations in terms of Clause 51. The Engineer may use a revised drawing or specification to cover several variations. It is up to the Contractor to make the case that any Clause 7 issue constitutes several ordered variations where it is relevant. This may be important in a claim which is based on, say, the sheer number of variations and the consequences of their cumulative delaying effect. The Contractor would be wise to heed and employ this right where necessary, bearing in mind the requirement to give *adequate* notice. In the case where the Contractor submits a programme which contains an early completion date or where the rate of progress is such that completion will be effected early then the courts have established that the Contractor has a right to complete early but there is no implied term requiring the Employer or his agents (including the Engineer) to perform their own obligations earlier in order to facilitate early completion by the Contractor (*Glenlion Construction v. Guinness Trust* (1988) 39 BLR 89).

Sub-Clause (3) entitles the Contractor to redress in the form of an award of an extension of time (pursuant to Clause 44) and payment of reasonable cost incurred (pursuant to Clause 52(4) which relates to notice of claims and Clause 60 which deals with monthly statements).

The *5th Edition of the ICE Conditions of Contract* uses the word 'deem' or 'deemed' in several Clauses and means that a particular action is assumed to have been taken regardless of whether it actually has or not. So, in this case, the issue of drawings or specifications is treated as if it was done pursuant to Clause 51 which relates to variations which are ordered by the Engineer.

[Contractor's address]

[Engineer]

[Date]

Dear Sir

**[Contract description]**
**[Contract location]**
**[Area of work for which drawings and/or instructions are sought]**
**Request for further [drawings and/or instructions]**

We write to request the issue of [drawings/instructions] in order to allow us to commence/complete the construction of [          ]. In order to avoid delay and extra cost this information is required by [          ].

This letter constitutes a notice issued pursuant to Clause 7(2) of the Conditions of Contract.

Yours faithfully


C Berry
Agent
for Unlimited Contracting Ltd


*Standard letter 1. Clause 7(2). Request for further drawings and/or instructions*

[Contractor's address]

[Engineer]

[Date]

Dear Sir

**[Contract description]**
**[Contract location]**
**[Area of work for which drawings and/or instructions are sought]**
**Notice of delay in the issue of documents**

We refer to our letter to you dated [          ] requesting the issue of further drawings/instructions related to the above, such request being made pursuant to Clause 7(2) of the Conditions of Contract.

We write to advise you that [as of the above date we have received no response thereto/the information was not supplied until [          ]] and as a consequence has delayed the execution of the Works by [          ] [days/weeks]. Extra cost in the form of [          ] has been incurred.

We hereby request that pursuant to Clause 7(3) of the Conditions of Contract you grant an extension of time under the terms of Clause 44(1) of the Conditions of Contract and certify payment of this extra cost under the terms of Clause 60 of the Conditions of Contract.

We shall include a sum related hereto in the next monthly statement submitted in accordance with Clause 60(1) and confirm that appropriate contemporary records have been kept to support this application for an extension of time and additional payment. Please note that this letter constitutes a notice required pursuant to Clause 52(4)(b) of the Conditions of Contract.

Yours faithfully


B Diddley
Agent
for Unlimited Contracting Ltd

# THE 5TH EDITION OF THE ICE CONDITIONS OF CONTRACT | 23

*Standard letter 2. Clause 7(3). Notice of delay and extra cost as a result of a delay in the issue of drawings/instructions*

---

Standard letter 2 illustrates several significant principles. It is issued primarily to persuade the Engineer to exercise his powers under Sub-Clause (3) by granting an extension of time and certifying payment of cost to the Contractor. Adopting this approach ensures that the Engineer is made aware that the Contractor has experienced delay and incurred extra cost. It records this view on a contemporary basis when recollections by all parties are most likely to be at their most accurate. Part of the wording of Sub-Clause (3) is, as will be seen, common to a number of Clauses:

> If ... the Contractor suffers delay or incurs cost then the Engineer shall take such delay into account in determining any extension of time to which the Contractor is entitled under Clause 44 and the Contractor shall subject to Clause 52(4) be paid in accordance with Clause 60 the amount of such cost as may be reasonable.

This confers the Engineer with the power to grant extensions of time and certify the payment of extra cost. It is *always* the Engineer who has this power. The Contractor cannot make a claim directly under this Clause, he has to do so via Clause 52(4). This is repeated throughout the *5th Edition of the ICE Conditions of Contract*. The Contractor is to be paid 'in accordance with Clause 60' which means that where the Engineer agrees to recompense the Contractor then the amount is to be included in interim and final certificates and is subject to retention all as defined in Clause 60.

Note that the Model Contract Document for Highway Works (Section 1, Parts 2–5 inclusive of Volume 0: 'Model Contract Document for Major Works and Implementation Requirements') adds the word 'specifications' after drawings and instructions in this Clause.

## 2.5. Clause 11

### SUB-CLAUSE (1)—INSPECTION OF SITE

> The Contractor shall be deemed to have inspected and examined the Site and its surroundings and to have satisfied himself before submitting his tender as to the nature of the ground and sub-soil (so far as is

practicable and having taken into account any information in connection therewith which may have been provided by or on behalf of the Employer) the form and nature of the Site the extent and nature of the work and materials necessary for the completion of the Works the means of communication with and access to the Site the accommodation he may require and in general to have obtained for himself all necessary information (subject as above-mentioned) as to risks contingencies and all other circumstances influencing or affecting his tender.

### SUB-CLAUSE (2)—SUFFICIENCY OF TENDER

The Contractor shall be deemed to have satisfied himself before submitting his tender as to the correctness and sufficiency of the rates and prices stated by him in the Priced Bill of Quantities which shall (except in so far as it is otherwise provided in the Contract) cover all his obligations under the Contract.

Sub-Clause (1) deems the Contractor to have taken certain steps before setting about completing his tender. The *5th Edition of the ICE Conditions of Contract* uses the word 'deem' or 'deemed' so whether he did so or not interpretation of the contract will be based on an assumption that the action described was undertaken. In the case of this Clause, the contractor will have examined the site etc. and have satisfied himself as to the nature of the ground and subsoil. He will have taken into account any information supplied by the Employer such as a site investigation report, the extent and nature of the Works, the means of communication to the site, accommodation needs, all necessary information relating to risks and contingencies and all other circumstances. The list seems fairly extensive and therefore onerous for the Contractor but he can only take account of these factors to the extent that he can reasonably do so bearing in mind any limitations connected with the information provided. For example, if the site investigation was in error then this Clause could not be cited by the Engineer as a reasonable defence against an earthworks claim if delays and/or extra costs were incurred.

Sub-Clause (2) underlines the Contractor's common law obligation to stick to the rates in his tender (since they form part of the contract) assuming that there are no circumstances which warrant change. If the Contractor, say, forgets to include the surfacing element in his rate for the footway item then he has no redress. The Engineer, even if minded to do so, would have no power to make any allowance for this error; to do so would be contrary to the interests of the Employer.

Claims emanating under this Clause are normally made pursuant to Clause 12, as discussed later. It is suggested that it may be possible to thwart the tendency of the Engineer to repudiate a Clause 12 claim by referring to the Contractor's obligations under Clause 11 by alluding to Clause 11 in related correspondence (see Standard letter 3).

## 2.6. Clause 12

### SUB-CLAUSE (1)—ADVERSE PHYSICAL CONDITIONS AND ARTIFICIAL OBSTRUCTIONS

If during the execution of the Works the Contractor shall encounter physical conditions (other than weather conditions or conditions due to weather conditions) or artificial obstructions which conditions or obstructions he considers could not reasonably have been foreseen by an experienced contractor and the Contractor is of opinion that additional cost will be incurred which would not have been incurred if the physical conditions or artificial obstructions had not been encountered he shall if he intends to make any claim for additional payment give notice to the Engineer pursuant to Clause 52(4) and shall specify in such notice the physical conditions and/or artificial obstructions encountered and with the notice if practicable or as soon as possible thereafter give details of the anticipated effects thereof the measures he is taking or is proposing to take and the extent of the anticipated delay in or interference with the execution of the Works.

### SUB-CLAUSE (2)—MEASURES TO BE TAKEN

Following receipt of a notice under sub-clause (1) of this Clause the Engineer may if he thinks fit inter alia

(a) require the Contractor to provide an estimate of the cost of the measures he is taking or is proposing to take

(b) approve in writing such measures with or without modification

(c) give written instructions as to how the physical conditions or artificial obstructions are to be dealt with

(d) order a suspension under Clause 40 or a variation under Clause 51.

### SUB-CLAUSE (3)—DELAY AND EXTRA COST

To the extent that the Engineer shall decide that the whole or some part of the said physical conditions or artificial obstructions could not reasonably have been foreseen by an experienced contractor the Engineer shall take any delay suffered by the Contractor as a result of such conditions or obstructions into account in determining any extension of time to which the Contractor is entitled under Clause 44 and the Contractor shall subject to Clause 52(4) (notwithstanding that the Engineer may not have given any instructions or orders pursuant to sub-clause (2) of this Clause) be paid in accordance with Clause 60 such sum as represents the reasonable cost of carrying out any additional work done and additional Constructional Plant used which would not have been done or used had such conditions or obstructions or such part thereof as the case may be not been encountered together with a reasonable percentage addition thereto in respect of profit and the reasonable costs incurred by the Contractor by reason of any unavoidable delay or disruption of working suffered as a consequence of encountering the said conditions or obstructions or such part thereof.

### SUB-CLAUSE (4)—CONDITIONS REASONABLY FORESEEABLE

If the Engineer shall decide that the physical conditions or artificial obstructions could in whole or in part have been reasonably foreseen by an experienced contractor he shall so inform the Contractor in writing as soon as he shall have reached that decision but the value of any variation previously ordered by him pursuant to sub-clause (2)(d) of this Clause shall be ascertained in accordance with Clause 52 and included in the Contract Price.

Sub-Clause (1) requires the Contractor to give written notice to the Engineer if the Contractor

(a) encounters physical conditions or artificial obstructions
(b) believes that these could not have reasonably been foreseen by an experienced Contractor (irrelevant but interesting to note that it need not be the Contractor carrying out the Works)
(c) will incur additional cost; and
(d) intends to make a claim under Clause 52(4) (Notice of Claims).

Note that there is no timescale given but presumably it is subject to the test of reasonableness in all the circumstances. The notice should specify the physical conditions or artificial obstructions encountered and should with the notice or as soon as practicable (which means, arguably, when it suits the Contractor again subject to the test of reasonableness) give details of three matters

(i) the anticipated effects of this event

(ii) the measures taken or proposed
(iii) the amount of the delay or disruption.

Sub-Clause (2) entitles the Engineer to require the Contractor to provide an estimate (not a quotation) of the cost of the actual or proposed measures, approve measures in writing, give written instructions on how the adverse physical conditions or artificial obstructions may be dealt with or order a suspension pursuant to Clause 40 or a variation pursuant to Clause 51. Again no timescale is mentioned but the doctrine of reasonableness in all the circumstances would presumably apply. Note that these are all powers which he may choose to exercise—they are not obligatory. Note also the use of the catch-all phrase *'inter alia'*—amongst other things. Theoretically, therefore, the Engineer can take any other steps which he deems necessary. The freedom of the Engineer in such circumstances is frequently limited; the Contractor is the specialist in construction matters and almost always has the edge in controlling events in this situation. For example, how many Engineers would feel confident about suggesting that a particular excavator has a capacity which exceeds the needs of the situation? In dayworks operations, plant is generally paid for on the basis of its 'maker's rated nominal weight of machine' and so the greater its weight the higher the rate paid. It may be in the Contractor's financial interests to employ plant which is rated at a higher value than the situation truly warrants.

Sub-Clause (3) permits the Engineer to award the Contractor an extension of time pursuant to Clause 44 and certify payment of his costs in accordance with Clause 60 ('Monthly Statements') together with a reasonable percentage in addition thereto for profit 'in respect of delay and disruption', a very famous phrase in contracting circles but note that such payments are to the extent that the Engineer thinks that the 'whole or some part of the said physical conditions or artificial obstructions could not reasonably have been foreseen by an experienced contractor'. As ever, the Contractor has no right of claim under this Clause and would have to seek redress via Clause 52(4)(a) or (b), as appropriate (Clause 52(4)(a) if a variation is ordered or Clause 52(4)(b) if no variation is ordered).

Sub-Clause (4) requires the Engineer to write to the Contractor if it is considered that all or part of the 'physical conditions or artificial obstructions could in whole or in part have been reasonably foreseen by an experienced contractor' and the value of any ordered variation ascertained in accordance with Clause 52 ('Valuation of Ordered Variations').

Examples of unforeseen physical conditions would be soils which contain more moisture or have a higher proportion of rock than is suggested by the site investigation report (or other contract documents). Examples of artificial obstructions would be a drain line falling outside of the boundary within which the Works are supposed to fall. Note that weather conditions or the results of weather conditions are excluded under this Clause. Often, weather conditions and their effects attract an extension of time but without the benefit of extra costs being paid to the Contractor. Contractors should also note the need to maintain contemporary records, a requirement of Clause 52(4). It falls to the Contractor to prove that a condition or obstruction is not reasonably foreseeable to an experienced contractor.

The phrase 'could not reasonably have been foreseen by an experienced contractor' warrants careful consideration. It defines what constitutes an adverse physical condition or artificial obstruction but is ambiguous in that it does not specify the level of probability at which a claim becomes valid. Hudson[3] encapsulates the key question clearly and succinctly. In order to qualify for extension of time and payment of cost under this Clause, a condition (or obstruction) has to be not reasonably foreseeable. He poses the question, does not reasonably foreseeable mean 'possible though unlikely' or does it mean 'probable and likely'? In order to get an answer two other texts are helpful. Furmston[4] suggests that the Clause should be read as if it said 'foreseen as likely'. In other words unless the condition or obstruction is 'foreseen as likely' then it qualifies for extension of time and payment. This view is supported by Abrahamson[5] who suggests 'that a claim is barred only if an experienced contractor could have foreseen a substantial risk'. Note the use of the word substantial.

What about the quantum, how much can the Contractor claim if it can be established that a condition or obstruction is not reasonably foreseeable? In the case of the work involved in dealing with the problem then he is entitled to 'such sum as represents the reasonable cost' and 'a reasonable percentage addition thereto in respect of profit'. Where there is an element of 'delay or disruption' then the Contractor is entitled to only 'the reasonable cost'.

The procedure which the Contractor should follow is given in standard letters 3 to 5. It should be noted in these instances particularly that the letters employ much wording from the Clause itself, the wisdom of which has been discussed earlier.

This is a very important Clause as far as Contractors are concerned. It is probably the Clause which is most often invoked in claims (apart from Clause 52(4), of course). Contractors should issue such notices as soon as the possibility of extra cost arises; they need not necessarily be pursued subsequently.

There is a tendency nowadays for persons framing contract documents to write out Clause 12 in the belief that it frees the Employer of any responsibility in terms of unforeseen adverse physical conditions or artificial obstructions. This is not the case, as a Contractor would still have recourse to common law provisions related to matters which were unforeseen at the time of tender, and, hence, writing out Clause 12 may have no effect. Indeed, the effect may be to add to the Employer's costs since the Contractor may have to revert to court action in order to secure payment.

There remains a wide disparity of views (this is often the case when seeking legal advice) on the success of repudiating claims on the basis of them being time-barred. Say, for example, a Contractor's first intimation of a claim under Clause 12 is its inclusion in his final measure. Such an approach would appear to be clearly contrary to the requirements of the Clause. It is possible that the courts may support repudiation on the basis that the necessary notice was not given. It may be that the courts would decide that the failure of the Contractor to give notice, although a clear breach of the terms of Clause 12, did not affect the basic merit of the claim nor did it prejudice the Engineer's ability to adjudge its merits on a contemporaneous basis. The Contractor takes an unnecessary risk when he fails to serve notice. He need not pursue the claim if subsequent events dictate such an approach but serving notice avoids the possibility of repudiation on the basis of not complying with a condition precedent if notice is served. It is always a sound approach to comply with the Clause even if the possibility of later pursuit is remote. Losing a claim on such a technicality is commercially inexcusable.

Some texts have complained about Clause 12 on the basis that it is an open door for claims. The way that it has been framed (reflecting the entire approach of the *5th Edition of the ICE Conditions of Contract*) is to allow the Contractor to give his most competitive price (or most economically advantageous tender as it is often currently fashionably called) without the need to allow a sum of money for unforeseen conditions etc. Those who let contracts cannot have the benefit of both the cheapest possible tender and expect the Contractor to take the risk. If the Employer expects the Contractor to take such risks then that should be made clear in the contract documentation but it is more than likely that the resulting tenders will be measurably higher consistent with the degree of risk. It is far more cost-effective to invest some time, effort, thought and cost in carrying out a proper site investigation.

[Contractor's address]

[Engineer]

[Date]

Dear Sir

**[Contract description]**
**[Contract location]**
**[Area of work affected by adverse physical condition or artificial obstruction]**
**Notice of encountering adverse [physical condition/artificial obstruction]**

We have to advise you that we have encountered an adverse [physical condition/artificial obstruction] in the form of [         ] at [         ]. In our opinion this could not reasonably have been foreseen by an experienced Contractor bearing in mind our obligations under Clause 11(1) and the information supplied in relation thereto. The effect of this adverse [physical condition/artificial obstruction] is that we shall incur additional cost which would not have been incurred if this [condition/obstruction] had not been encountered.

In order to deal with the problem we are [detail measures proposed/being taken]. Details of any delay or disruption will be sent to you in due course.

This notice is issued pursuant to Clause 12(1) and Clause 52(4)(b) of the Conditions of Contract.

Yours faithfully


R Orbison
Agent
for Unlimited Contracting Ltd


*Standard letter 3. Clause 12(1). Notice of encountering adverse physical condition/artificial obstruction and measures proposed/being taken*

[Contractor's address]

[Engineer]

[Date]

Dear Sir

**[Contract description]**
**[Contract location]**
**[Area of work affected by adverse physical condition or artificial obstruction]**
**Estimate of costs for dealing with adverse physical condition or artificial obstruction**

Further to your letter dated [           ], please find attached an estimate of the measures which we are [taking/proposing to take] to deal with the adverse [physical condition/artificial obstruction] detailed in our letter dated [           ].

This estimate is provided pursuant to Clause 12(2) of the Conditions of Contract.

Yours faithfully


S Cropper
Agent
for Unlimited Contracting Ltd

Enc

*Standard letter 4. Clause 12(2). Provision of ordered estimate of costs of actual or proposed measures to deal with adverse physical condition/artificial obstruction*

[Contractor's address]

[Engineer]

[Date]

Dear Sir

**[Contract description]**
**[Contract location]**
**[Area of work affected by adverse physical condition or artificial obstruction]**
**Notice of claim due to adverse physical condition/artificial obstruction**

In our letter to you dated [            ] we advised you that the adverse [physical condition/artificial obstruction] in the form of [            ] has resulted in us incurring additional cost which would not have been incurred if this [condition/obstruction] had not been encountered. Furthermore this has delayed and disrupted the execution of the Works. The amount of the delay is [            ] [days/weeks].

We hereby request that pursuant to Clause 12(3) of the Conditions of Contract you grant an extension of time under the terms of Clause 44(1) of the Conditions of Contract and certify payment of this extra cost under the terms of Clause 60 of the Conditions of Contract.

We shall include a sum related hereto in the next monthly statement submitted in accordance with Clause 60(1) and confirm that appropriate contemporary records have been kept to support this application for an extension of time and additional payment. Please note that this letter constitutes a notice required pursuant to Clause 52(4)(b) of the Conditions of Contract.

Yours faithfully


A King
Agent
for Unlimited Contracting Ltd

*Standard letter 5. Clause 12(3). Notice of claim as a result of encountering adverse physical condition/artificial obstruction*

---

Standard letters 3., 4. and 5. illustrate the form of notices issued in relation to Clause 12. The notice of intention to make a claim for additional payment required under Clause 12(1) or for the prompting letter for additional payment and extension of time issued pursuant to Clause 12(3) may be combined with the notice of unforeseen physical conditions or artificial obstructions issued under Clause 12(1). Similarly, the indication of anticipated effects of the condition or obstruction, the measures the Contractor has taken or is proposing to take, their estimated cost and the extent of the anticipated delay in or interference with the execution of the Works can be included in the initial letter, all depending on events and timescale.

## 2.7.   Clause 13

### SUB-CLAUSE (1)—WORK TO BE TO SATISFACTION OF ENGINEER

Save in so far as it is legally or physically impossible the Contractor shall construct complete and maintain the Works in strict accordance with the Contract to the satisfaction of the Engineer and shall comply with and adhere strictly to the Engineer's instructions and directions on any matter connected therewith (whether mentioned in the Contract or not). The Contractor shall take instructions and directions only from the Engineer or (subject to the limitations referred to in Clause 2) from the Engineer's Representative.

### SUB-CLAUSE (2)—MODE AND MANNER OF CONSTRUCTION

The whole of the materials plant and labour to be provided by the Contractor under Clause 8 and the mode manner and speed of construction and maintenance of the Works are to be of a kind and conducted in a manner approved of by the Engineer.

### SUB-CLAUSE (3)—DELAY AND EXTRA COST

If in pursuance of Clause 5 or sub-clause (1) of this Clause the Engineer shall issue instructions or directions which involve the Contractor in delay or disrupt his arrangements or methods of construction so as to cause him to incur cost beyond that reasonably to have been foreseen by an experienced contractor at the time of tender then the Engineer shall take such delay into account in determining any extension of time to which the Contractor is entitled under Clause 44 and the Contractor shall subject to Clause 52(4) be paid in accordance with Clause 60 the amount of such cost as may be reasonable. If such instructions or directions require any variation to any part of the Works the same shall be deemed to have been given pursuant to Clause 51.

To the extent that any requirement of the contract is not illegal or impossible, Sub-Clause (1) requires the Contractor to complete the Works in accordance with the contract and with any instructions issued by the Engineer or the Engineer's Representative. The latter is qualified in so far as any instructions should relate to the Works. The Engineer cannot order work which is unconnected with the Works (see Clause 51). The concept of impossible compliance should be clearly understood. If a method specification applies to a particular element of work and the Contractor correctly employs the method but the desired result is not achieved then the Contractor is not liable. Where a results specification is employed then the mere fact that it is extremely difficult and/or very expensive to execute does not make it impossible. An example would be where the contract calls for the ends of the carriageway to be tied into existing roads with a finished road level of ±6 mm. An Engineer who insists on achieving this specified compliance ignoring the fact that the road has acceptable rideability characteristics may well be asking for what is practically impossible. Proving it, however, is not easy.

It is important that the Contractor is mindful of the extent of the powers of the Engineer's Representative. If the Contractor obeys an instruction from the Engineer's Representative which is issued *ultra vires*, i.e. beyond his powers since it was issued outwith the scope of the powers delegated to the Engineer's Representative in accordance with Clause 2(2), and this instruction is not subsequently supported by the Engineer then the Contractor is unlikely to get support from the courts. Having said that, the Engineer is morally obliged to support such instructions although his primary duty of care to the Employer may prevent this.

Where the Contractor obeys an invalid instruction from the Engineer's Representative then such a situation would not produce a valid claim. Where the Engineer does not support an invalid instruction from the Engineer's Representative then that situation

# THE 5TH EDITION OF THE ICE CONDITIONS OF CONTRACT | 35

would, again, of itself, not provide grounds for a valid claim.

Sub-Clause (2) requires the Contractor to use materials, plant and labour which are approved by the Engineer. The mode, manner and speed of construction along with the maintenance of the Works are also subject to the same assent. Of course, such approval is set down in the contract in the specification and other contract documents. Where the Engineer is unhappy with any element of the Works this disagreement must be based on some element of the specification or other contractual requirement and not merely a notional opinion e.g. condemning chipped hot rolled asphalt wearing course on the basis that, in the Engineer's opinion, too many of the chippings have failed to remain embedded in the wearing course. Where the Engineer condemns a piece of work and justification for this under the specification is uncertain, the Contractor should request clarification of the contract element which has been breached. The Engineer would be wise to forestall any such criticism by citing the Clause in the specification or other contract document wherein the requirements have not been met in any communication which condemns any piece of work.

Sub-Clause (3) entitles the Engineer to recognise, by means of the award of extension of time and cost (not profit as in Clause 12(3) for additional work), that instructions issued pursuant to Clause 5 (clarifying ambiguities or discrepancies in the documents) or pursuant to Sub-Clause (1) of this Clause (instructions and directions issued by the Engineer in connection with the construction completion and maintenance of the Works) caused the Contractor to encounter delay or disruption. It removes the need for an ordered variation by deeming any such instruction or direction to have been issued pursuant to Clause 51 ('ordered variations').

Where the Engineer does not recognise that a Contractor has been delayed or disrupted then the Contractor should issue a notice under Clause 52(4)(a) or (b) (Clause 52(4)(a) if a variation is ordered or Clause 52(4)(b) if no variation is ordered).

Where circumstances permit then the Contractor should pursue delay and disruption claims under Clause 12 rather than Clause 13 because the former will give a profit element whereas the latter will not.

Certain situations occur in construction works where the Contractor feels he has encountered extra costs above those which could reasonably have been expected at the time of tender on the basis of the contract documentation but has the difficulty of not having a contract condition against which to register a claim. In such circumstances, of course, it could be pursued on the basis of a common law breach but often it is preferable to cite the matter under a specific Clause. This Clause may come to his aid if the Contractor requests the Engineer to issue an instruction in relation to the matter in hand and thus, the resultant instruction would fall within the sphere of this Clause.

Again that famous Contractor's phrase appears in this Clause—'delay and disruption'.

---

[Contractor's address]

[Engineer]

[Date]

Dear Sir

**[Contract description]**
**[Contract location]**
**[Subject of instruction]**
**Notice of delay and/or disruption involving additional cost as a result of Engineer's instruction**

We refer to your instruction dated [         ] related to [         ] and issued pursuant to [Clause 5/Clause 13(1)] of the Conditions of Contract.

This instruction has delayed and/or disrupted the execution of the Works and has resulted in us incurring cost beyond that reasonably to have been foreseen by an experienced contractor at the time of tender. The amount of the delay is [         ] [days/weeks].

We hereby request that pursuant to Clause 13(3) of the Conditions of Contract you grant an extension of time under the terms of Clause 44(1) of the Conditions of Contract and certify payment of this extra cost under the terms of Clause 60 of the Conditions of Contract.

We shall include a sum related hereto in the next monthly statement submitted in accordance with Clause 60(1) and confirm that appropriate contemporary records have been kept to support this application for an extension of time and additional payment. Please note that this letter constitutes a notice required pursuant to Clause 52(4)(b) of the Conditions of Contract.

Yours faithfully


B White
Agent
for Unlimited Contracting Ltd


*Standard letter 6. Clause 5/13(1). Notification of claim as a result of delay and/or disruption resulting from the Engineer's instruction*

## 2.8. Clause 14

### SUB-CLAUSE (1)—PROGRAMME TO BE FURNISHED

Within 21 days after the acceptance of his Tender the Contractor shall submit to the Engineer for his approval a programme showing the order of procedure in which he proposes to carry out the Work and thereafter shall furnish such further details and information as the Engineer may reasonably require in regard thereto. The Contractor shall at the same time also provide in writing for the information of the Engineer a general description of the arrangements and methods of construction which the Contractor proposes to adopt for the carrying out of the Works.

### SUB-CLAUSE (2)—REVISION OF PROGRAMME

Should it appear to the Engineer at any time that the actual progress of the Works does not conform with the approved programme referred to in sub-clause (1) of this Clause the Engineer shall be entitled to require the Contractor to produce a revised programme showing the modifications to the original programme necessary to ensure completion of the Works or any Section within the time for completion as defined in Clause 43 or extended time granted pursuant to Clause 44(2).

### SUB-CLAUSE (3)—METHODS OF CONSTRUCTION

If requested by the Engineer the Contractor shall submit at such times and in such detail as the Engineer may reasonably require such information pertaining to the methods of construction (including Temporary

Works and the use of Constructional Plant) which the Contractor proposes to adopt or use and such calculations of stresses strains and deflections that will arise in the Permanent Works or any parts thereof during construction from the use of such methods as will enable the Engineer to decide whether if these methods are adhered to the Works can be executed in accordance with the Drawings and Specification and without detriment to the Permanent Works when completed.

### SUB-CLAUSE (4)—ENGINEER'S CONSENT

The Engineer shall inform the Contractor in writing within a reasonable period after receipt of the information submitted in accordance with sub-clause (3) of this Clause either

(a) that the Contractor's proposed methods have the consent of the Engineer

(b) in what respects in the opinion of the Engineer they fail to meet the requirements of the Drawings or Specification or will be detrimental to the Permanent Works.

In the latter event the Contractor shall take such steps or make such changes in the said methods as may be necessary to meet the Engineer's requirements and to obtain his consent. The Contractor shall not change the methods which have received the Engineer's consent without the further consent in writing of the Engineer which shall not be unreasonably withheld.

### SUB-CLAUSE (5)—DESIGN CRITERIA

The Engineer shall provide to the Contractor such design criteria relevant to the Permanent Works or any Temporary Works designed by the Engineer as may be necessary to enable the Contractor to comply with sub-clauses (3) and (4) of this Clause.

### SUB-CLAUSE (6)—DELAY AND EXTRA COST

If the Engineer's consent to the proposed methods of construction shall be unreasonably delayed or if the requirements of the Engineer pursuant to sub-clause (4) of this Clause or any limitations imposed by any of the design criteria supplied by the Engineer pursuant to sub-clause (5) of this Clause could not reasonably have been foreseen by an experienced contractor at the time of tender and if in consequence of any of the aforesaid the Contractor unavoidably incurs delay or cost the Engineer shall take such delay into account in determining any extension of time to which the Contractor is entitled under Clause 44 and the Contractor shall subject to Clause 52(4) be paid in accordance with Clause 60 such sum in respect of the cost incurred as the Engineer considers fair in all the circumstances.

### SUB-CLAUSE (7)—RESPONSIBILITY UNAFFECTED BY APPROVAL

Approval by the Engineer of the Contractor's programme in accordance with sub-clauses (1) and (2) of this Clause and the consent of the Engineer to the Contractor's proposed methods of construction in accordance with sub-clause (4) of this Clause shall not relieve the Contractor of any of his duties or responsibilities under the Contract.

The requirements of this Clause are largely self-explanatory but the provision of a detailed programme by the Contractor will enable the demonstration of the delaying and disrupting effects of any events on site which cause delay and disruption and it is likely that arguments will find more support with the Engineer if a programme has been established at the outset of the contract. Although there is no contractual need for the Contractor to attach time periods to the programme, it would be wise so to do, in case the programme becomes evidential during claims discussions. Notwithstanding, the prudent Engineer will require the Contractor to justify his programme if there are any doubts about the ability of the Contractor to meet the timescales set out therein. Again the Engineer is hampered by the fact that matters of construction are more likely to be the province of the Contractor.

The programme is not a contract document since it does not form part of the tender submission, unless of course, that is an express requirement of the contract established by an alteration to the standard *5th Edition of the ICE Conditions of Contract.*

Within 21 days of the award of the contract, the Contractor must submit a programme showing the order in which he proposes to carry out the Works. A general description of the arrangements and methods of construction which he proposes to adopt must also be provided.

According to Sub-Clause (2), the Engineer can require the Contractor to produce a revised programme if it appears to the Engineer that actual and planned progress do not match; the revised programme must show how completion will be effected in order to meet the time for completion. In contrast however, the Contractor cannot be required to provide a revised general description of the arrangements and methods of construction which he proposes to adopt.

Note the use of the words 'arrangements' and 'methods of construction' in both Clause 13(3) and Clause 14(1) thus providing a claims link between the two Clauses.

Sub-Clause (3) requires the Contractor to submit in such detail and at such frequencies as may be reasonably required by the Engineer, information pertaining to the methods of construction.

Within a reasonable period, according to Sub-Clause (4), after the receipt of the information specified in Sub-Clause (3), the Engineer shall write to the Contractor advising that the Contractor's proposed methods are approved or shall define in what respects they do not meet the requirements of the drawings or the specification or will be detrimental to the Permanent Works. If the Engineer lists deficiencies then the Contractor will take any steps necessary to get the approval of the Engineer i.e. meet the specification etc. Approved methods are not to be changed without the Engineer's consent.

Sub-Clause (5) requires the Engineer to provide design criteria necessary to comply with Sub-Clauses (3) and (4) of this Clause.

Sub-Clause (6) is the provision which permits the Engineer to recognise that the Contractor has incurred cost or been delayed because the Engineer unreasonably delays approval or because the requirements of the Engineer pursuant to Sub-Clause (4) hereof or the design criteria could not reasonably have been foreseen by an experienced contractor. In the absence of such recognition the Contractor could claim additional payment (pursuant, of course, to Clause 52(4)) and extension of time (pursuant to Clause 44).

Sub-Clause (7) establishes that the Contractor's liabilities under the contract are not reduced because the Engineer approves the programme.

Some Engineers neither give approval nor list defects, but simply acknowledge that it is a programme. They do so in order to withhold approval and thus, they believe, avoid claims based on the Engineer's approval of the programme. It is clear from the wording that Sub-Clause (1) places a duty on the Engineer to either approve or list defects in the programme. In the vast majority of cases, these defects will be (albeit eventually) eliminated to the satisfaction of the Engineer after which approval should be granted.

The Engineer would be wise to ensure that he has adequate control over the supply of programme information and once in receipt of this information, invest adequate effort to ensure that the programme will meet the contractual requirements in terms of achieving completion by the stated time without being over-optimistic. Actual progress should also be monitored against that suggested by the programme in case, say, instructions or variations result in a claim which is based, at least partially, on the submitted programme. However, engineers should bear in mind the outcome of *Yorkshire Water Authority v. (Sir Alfred) McAlpine Ltd* (1985) 32 BLR 114, particularly in view of Hudson's[6] comments. This judgement is discussed in more detail in Section 2.23 (Clause 51 variations).

This Clause illustrates an interesting legal convention related to time. The law does not recognise part of a day so that when a Clause says that following a specified event, a notice will be served within seven days then if the event happened on a Monday morning then the notice could be served up to the end of Tuesday (i.e. midnight) the following week. In specifying a time limit, the law takes the view that 'days' includes holidays, Saturdays and Sundays unless a phrase such as 'working days' is employed. Months are treated as calendar months and not lunar months so, for example, an event which requires notice within three months and which occurred on 5 November will be met legally if the notice is served before midnight on 5 February.

---

[Contractor's address]

[Engineer]

[Date]

Dear Sir

**[Contract description]**
**[Contract location]**
**Notice of delay and extra cost as a result of delay in approval of Contractor's proposed method of construction**

We refer to our letter to you dated [         ] which contained the programme which showed the proposed order of procedure of the Works issued pursuant to Clause 14(1) of the Conditions of Contract. [As of the above date we have received no response thereto/consent was not granted until [         ]] and as a consequence delay and extra cost have been incurred.

We hereby request that pursuant to Clause 14(6) of the Conditions of Contract you grant an extension of time under the terms of Clause 44(1) of the Conditions of Contract and certify payment of this extra cost under the terms of Clause 60 of the Conditions of Contract.

We shall include a sum related hereto in the next monthly statement submitted in accordance with Clause 60(1) and confirm that appropriate contemporary records have been kept to support this application for an extension of time and additional payment. Please note that this letter constitutes a notice required pursuant to Clause 52(4)(b) of the Conditions of Contract.

Yours faithfully


S Wonder
Agent
for Unlimited Contracting Ltd


*Standard letter 7. Clause 14(6). Notice of claim as a result of an unreasonable delay in approval of proposed order of procedure*

---

[Contractor's address]

[Engineer]

[Date]

Dear Sir

**[Contract description]**
**[Contract location]**
**Notice of delay and extra cost as a result of delay in approval of Contractor's proposed method of construction**

We refer to your letter dated [            ] which detailed the criteria relevant to the design of [item of Permanent Works or Temporary Works] issued pursuant to Clause 14(5) of the Conditions of

# THE 5TH EDITION OF THE ICE CONDITIONS OF CONTRACT | 43

Contract. These criteria are more onerous than those set in [relevant part] of the Design Manual for Road and Bridge Works and thus impose limitations which could not reasonably have been foreseen by an experienced contractor at the time of tender. As a result, we have unavoidably incurred a delay of [          ] [days/weeks] and extra cost in producing the design.

We hereby request that pursuant to Clause 14(6) of the Conditions of Contract you grant an extension of time under the terms of Clause 44(1) of the Conditions of Contract and certify payment of this extra cost under the terms of Clause 60 of the Conditions of Contract.

We shall include a sum related hereto in the next monthly statement submitted in accordance with Clause 60(1) and confirm that appropriate contemporary records have been kept to support this application for an extension of time and additional payment. Please note that this letter constitutes a notice required pursuant to Clause 52(4)(b) of the Conditions of Contract.

Yours faithfully


B Withers
Agent
for Unlimited Contracting Ltd


*Standard letter 8. Clause 14(6). Notice of claim due to limitations imposed by the provision of design criteria not reasonably foreseeable by an experienced contractor at the time of tender*

---

## 2.9.   Clause 17

### SETTING-OUT

The Contractor shall be responsible for the true and proper setting-out of the Works and for the correctness of the position levels dimensions and alignment of all parts of the Works and for the provision of all necessary instruments appliances and labour in connection therewith. If at any time during the progress of the Works any error shall appear or arise in the position levels dimensions or alignment of any part of the Works the Contractor on being required so to do by the Engineer shall at his own

## 44 | CLAIMS ON HIGHWAY CONTRACTS

cost rectify such error to the satisfaction of the Engineer unless such error is based on incorrect data supplied in writing by the Engineer or the Engineer's Representative in which case the cost of rectifying the same shall be borne by the Employer. The checking of any setting-out or of any line or level by the Engineer or the Engineer's Representative shall not in any way relieve the Contractor of his responsibility for the correctness thereof and the Contractor shall carefully protect and preserve all bench-marks sight rails pegs and other things used in setting out the Works.

---

This Sub-Clause establishes the Contractor's responsibility for the financial ramifications arising as a result of any error in setting out unless such error is due to incorrect data supplied in writing by the Engineer or the Engineer's Representative in which case the Contractor is entitled to an extra in the form of the cost of rectification. This Clause does not mention extension of time but Clause 44(1) ('Extension of time for completion') (see section 2.18 page 66), affords that right and would allow him to claim for an extension of time.

---

[Contractor's address]

[Engineer]

[Date]

Dear Sir

**[Contract description]**
**[Contract location]**
**[Subject of incorrect setting out data]**
**Notice of delay and extra cost as a result of incorrect setting out data**

We refer to the data supplied [in/on] [            ] related to the setting out of [            ]. This information has subsequently been shown to be incorrect. Delay and extra cost have been encountered as a result and we seek to recover all associated costs.

This letter serves as a notice required pursuant to Clause 52(4)(b) of the Conditions of Contract. Appropriate contemporary records will be kept to support an application for additional payment and extension of time pursuant to Clause 44(1) of the Conditions of Contract.

Yours faithfully


E Floyd
Agent
for Unlimited Contracting Ltd


*Standard letter 9. Clause 17(2). Notice of claim as a result of an error in the setting out data supplied by the Engineer or Engineer's Representative*

---

Note that standard letter 9. is one of the few where the right to recognise additional cost is not accorded to the Engineer and it would be left to the Contractor to claim under Clause 52(4)(b) for extra cost. A compensatory extension of time would be requested on the basis of the wording of Clause 44(1) 'other special circumstances of any kind whatsoever'.

## 2.10  Clause 20

### SUB-CLAUSE (1)—CARE OF THE WORKS

The Contractor shall take full responsibility for the care of the Works from the date of the commencement thereof until 14 days after the Engineer shall have issued a Certificate of Completion for the whole of the Works pursuant to Clause 48. Provided that if the Engineer shall issue a Certificate of Completion in respect of any Section or part of the Permanent Works before he shall issue a Certificate of Completion in respect of the whole of the Works the Contractor shall cease to be responsible for the care of that Section or part of the Permanent Works 14 days after the Engineer shall have issued the Certificate of Completion in respect of that Section or part and the responsibility for the care thereof shall thereupon pass to the Employer. Provided further that the Contractor shall take full responsibility for the care of any outstanding work which he shall have undertaken to finish during the Period of Maintenance until such outstanding work is complete.

### SUB-CLAUSE (2)—RESPONSIBILITY FOR REINSTATEMENT

In case any damage loss or injury from any cause whatsoever (save and except the Excepted Risks as defined in sub-clause (3) of this Clause) shall happen to the Works or any part thereof while the Contractor shall be responsible for the care thereof the Contractor shall at his own cost repair and make good the same so that at completion the Permanent Works shall be in good order and condition and in conformity in every respect with the requirements of the Contract and the Engineer's instructions. To the extent that any such damage loss or injury arises from any of the Excepted Risks the Contractor shall if required by the Engineer repair and make good the same as aforesaid at the expense of the Employer. The Contractor shall also be liable for any damage to the Works occasioned by him in the course of any operations carried out by him for the purpose of completing any outstanding work or of complying with his obligations under Clauses 49 and 50.

### SUB-CLAUSE (3)—EXCEPTED RISKS

The 'Excepted Risks' are riot war invasion act of foreign enemies hostilities (whether war be declared or not) civil war rebellion revolution insurrection or military or usurped power ionising radiations or contamination by radio-activity from any nuclear fuel or from any nuclear waste from the combustion of nuclear fuel radioactive toxic explosive or other hazardous properties of any explosive nuclear assembly or nuclear component thereof pressure waves caused by aircraft or other aerial devices travelling at sonic or supersonic speeds or a cause due to use or occupation by the Employer his agents servants or other contractors (not being employed by the Contractor) of any part of the Permanent Works or to fault defect error or omission in the design of the Works (other than a design provided by the Contractor pursuant to his obligations under the Contract).

Sub-Clause (1) places responsibility for the risk of accidental damage to the Works on the Contractor from the Date for Commencement to 14 days after the issue of the Certificate of Completion. Note that it is the issue and not the date of the Certificate which is relevant. The use of this term and the time lapse being specified as 14 days are probably to give the Employer time to take out appropriate insurance. Similar provisions apply where a section or substantial part are the subject of a Certificate of Completion (see

# THE 5TH EDITION OF THE ICE CONDITIONS OF CONTRACT | 47

Clause 48). The earlier the Certificate of Completion is issued the sooner the Contractor is freed of the need to insure against the risks embraced within this Clause.

Sub-Clause (2) makes the Contractor responsible for the costs of repairing and making good unless due to an 'excepted risk' as defined in Sub-Clause (3). Where the damage, loss or injury is due either wholly or in part to an excepted risk then the costs either wholly or in part are at the 'expense of the Employer'.

Sub-Clause (3) defines the excepted risks the meaning of which have very specific interpretations in the insurance world. When taking out insurance (see Clause 20 for contractual requirements) the Contractor is advised to consult a specialist and employ the wording of the Clause in the policy. It is suggested that the traditional wording of the *5th Edition of the ICE Conditions of Contract* does not list terrorism (in its entirety, according to professional insurance advice) as an excepted risk; perhaps that is the intention of the parties but the matter of responsibility should be given some careful thought and clarified where there is any doubt.

The issues which are most likely to be the subject of a dispute under excepted risks are 'a cause due to use or occupation by the Employer his agents servants or other contractors (not employed by the Contractor)' or 'fault defect error or omission in the design of the Works' other than design by the Contractor.

---

[Contractor's address]

[Engineer]

[Date]

Dear Sir

**[Contract description]**
**[Contract location]**
**[Details of damage/loss/injury]**
**Notice of delay and extra cost as a result of damage/loss/injury occurring arising from an excepted risk**

We refer to the [damage/loss/injury] which occurred at the above on [          ]. We believe that this is due to a matter covered by an excepted risk namely [            ] as defined in Clause 20(3) of the Conditions of Contract. In accordance with the terms of Clause 20(2) of the Conditions of Contract we seek expenses for the cost of [making good/repairing] the Works. Furthermore this work has caused a delay of [            ] [days/weeks].

This letter serves as a notice required pursuant to Clause 52(4)(b) of the Conditions of Contract. Appropriate contemporary records will be kept to support an application for additional payment and extension of time pursuant to Clause 44(1) of the Conditions of Contract.

Yours faithfully


B Holly
Agent
for Unlimited Contracting Ltd


*Standard letter 10. Clause 20(2). Notice of claim as a result of damage loss or injury occurring under an excepted risk*

## 2.11. Clause 27

### SUB-CLAUSE (1)—PUBLIC UTILITIES STREET WORKS ACT 1950: DEFINITIONS

For the purposes of this Clause

(a) the expression 'the Act' shall mean and include the Public Utilities Street Works Act 1950 and any statutory modification or re-enactment thereof for the time being in force

(b) all other expressions common to the Act and to this Clause shall have the same meaning as that assigned to them by the Act.

### SUB-CLAUSE (2)—NOTIFICATIONS BY EMPLOYER TO CONTRACTOR

The Employer shall before the commencement of the Works notify the Contractor in writing

(a) whether the Works or any parts thereof (and if so which parts) are Emergency Works

(b) which (if any) parts of the Works are to be carried out in Controlled Land or in a Prospectively Maintainable Highway.

If any duly authorised variation of the Works shall involve the execution thereof in a Street or in Controlled Land or in a Prospectively Maintainable Highway or are Emergency Works the Employer shall notify the Contractor in writing accordingly at the time such variation is ordered.

### SUB-CLAUSE (3)—SERVICE OF NOTICES BY EMPLOYER

The Employer shall (subject to the obligations of the Contractor under sub-clause (4) of this Clause) serve all such notices as may from time to time whether before or during the course of or after completion of the Works be required to be served under the Act.

### SUB-CLAUSE (4)—NOTICES BY CONTRACTOR TO EMPLOYER

The Contractor shall in relation to any part of the Works (other than Emergency Works) and subject to the compliance by the Employer with sub-clause (2) of this Clause give not less than 21 days' notice in writing to the Employer before

(a) commencing any part of the Works in a Street (as defined by Sections 1(3) and 38(1) of the Act)

(b) commencing any part of the Works in Controlled Land or in a Prospectively Maintainable Highway

(c) commencing in a Street or in Controlled Land or in a Prospectively Maintainable Highway any part of the Works which is likely to affect the apparatus of any Owning Undertaker (within the meaning of Section 26 of the Act).

Such notice shall state the date on which and the place at which the Contractor intends to commence the execution of the work referred to therein.

### SUB-CLAUSE (5)—FAILURE TO COMMENCE STREET WORKS

If the Contractor having given any such notice as is required by sub-clause (4) of this Clause shall not commence the part of the Works to which such notice relates within 2 months after the date when such notice is given such notice shall be treated as invalid and compliance with the said sub-clause (4) shall be requisite as if such notice had not been given.

## SUB-CLAUSE (6)—DELAYS ATTRIBUTABLE TO VARIATIONS

In the event of such a variation of the Works as is referred to in sub-clause (2) of this Clause being ordered by or on behalf of the Employer and resulting in delay in the execution of the Works by reason of the necessity of compliance by the Contractor with sub-clause (4) of this Clause the Engineer shall take such delay into account in determining any extension of time to which the Contractor is entitled under Clause 44 and the Contractor shall subject to Clause 52 be paid in accordance with Clause 60 such additional cost as the Engineer shall consider to have been reasonably attributable to such delay.

## SUB-CLAUSE (7)—CONTRACTOR TO COMPLY WITH OTHER OBLIGATIONS OF ACT

Except as otherwise provided by this Clause where in relation to the carrying out of the Works the Act imposes any requirement or obligations upon the Employer of the Contractor shall subject to Clause 49(5) comply with such requirements and obligations and shall (subject as aforesaid) indemnify the Employer against any liability which the Employer may incur in consequence of any failure to comply with the said requirements and obligations.

This Clause deals with the Public Utilities Street Works Act of 1950. This Act was superseded by the New Roads and Street Works Act 1991 and accordingly the terms of this Clause are applied to its replacement in accordance with the provisions set out in Sub-Clause (1)(a) and, therefore, the terms of the new Act apply herein.

Sub-Clause (2) requires the Employer to advise the Contractor in writing before the Commencement of the Works if any of the elements or the whole of the Works are emergency works as defined in the Act. The same notification requirement applies if any or all of the Works are in controlled land or in a prospectively maintainable highway. 'Emergency works', 'controlled land' and 'prospectively maintainable highway' are all terms from the Act.

Emergency works are defined in Section 52(1) as 'works whose execution at the time when they are executed is required in order to put an end to, or to prevent the occurrence of, circumstances then existing or imminent (or which the person responsible for the works believes on reasonable grounds to be existing or imminent) which are likely to cause damage to persons or property'.

Although 'controlled land' is not defined in the new Act, Section 162 has a marginal heading 'former controlled land'. It is suggested that controlled land is land which is adjacent to a road and is deemed to be controlled by the roads authority for the purpose of, say, future widening, falling within areas necessary for the provision of adequate sightlines etc. It is further suggested that the roads authority would have directed undertakers to place apparatus there for Section 162 to apply.

Prospectively maintainable highway is now replaced by the phrase 'prospective public road' which is referred to in Section 146 of the Act as 'a road in the [local roads authority] area which is ... likely to become a public road'.

Sub-Clause (2) continues that if any ordered variation is emergency works or is in controlled land or on a prospectively maintainable highway then the same written notification by the Employer is required.

Sub-Clause (3) specifies that notices required under the Act should continue to be served by the Employer.

Under Sub-Clause (4) the Contractor must give to the Employer at least 21 days' notice before commencing any Works (except emergency works) which are in a street, in controlled land or on a prospectively maintainable highway, all as defined in the Act. The same applies where the apparatus of an owning undertaker is likely to be affected in a street, in controlled land or on a prospectively maintainable highway. An 'undertaker' is defined in the Act in Section 107(4) as a person who has a statutory right to place apparatus in roads or person who has permission to do under Section 109 (person given permission by the roads authority to place apparatus in a road).

Sub-Clause (5) effectively cancels the notice in Sub-Clause (4) if the Contractor does not commence the works within two months of having given notice of intention to do so.

Sub-Clause (6) entitles the Engineer to take into account the fact that the Contractor has been delayed by granting an extension of time pursuant to Clause 44 and paying for the variation pursuant to the terms of Clause 52 in accordance with Clause 60. Note that it is the Employer who features within this Clause until Sub-Clause (6) where the Engineer has the task of granting an extension of time and recognising additional cost.

Most trunk roads are largely free of undertakers' apparatus and so it is unlikely that these provisions will feature in any contracts involving trunk roads.

## 52 | CLAIMS ON HIGHWAY CONTRACTS

[Contractor's address]

[Engineer]

[Date]

Dear Sir

**[Contract description]**
**[Contract location]**
**[Specify nature and location of variation]**
**Notice of delay and additional cost as a result of a variation related to the New Roads and Street Works Act**

We write to record that as a result of complying with the variation notified to us by the Employer pursuant to Clause 27(2) of the Conditions of Contract in his letter dated [          ], delay has been incurred and accordingly we request that you grant extension of time and certify payment of all additional costs attributable thereto pursuant to Clause 27(6) of the Conditions of Contract.

Please note that this letter constitutes a notice required pursuant to Clause 52(4)(b) of the Conditions of Contract. Appropriate contemporary records will be kept to support an application for additional payment and extension of time pursuant to Clause 44(1) of the Conditions of Contract.

Yours faithfully


D P Everly
Agent
for Unlimited Contracting Ltd




*Standard letter 11. Clause 27(6). Notice of claim as a result of compliance with a variation related to utility operations*

## 2.12. Clause 31

### SUB-CLAUSE (1)—FACILITIES FOR OTHER CONTRACTORS

The Contractor shall in accordance with the requirements of the Engineer afford all reasonable facilities for any other contractors employed by the Employer and their workmen and for the workmen of the Employer and of any other properly authorised authorities or statutory bodies who may be employed in the execution on or near the Site of any work not in the Contract or of any contract which the Employer may enter into in connection with or ancillary to the Works.

### SUB-CLAUSE (2)—DELAY AND EXTRA COST

If compliance with sub-clause (1) of this Clause shall involve the Contractor in delay or cost beyond that reasonably to be foreseen by an experienced contractor at the time of tender then the Engineer shall take such delay into account in determining any extension of time to which the Contractor is entitled under Clause 44 and the Contractor shall subject to Clause 52(4) be paid in accordance with Clause 60 the amount of such cost as may be reasonable.

Under Sub-Clause (1), the Contractor is required, in accordance with the requirements of the Engineer, to 'afford all reasonable facilities for any other contractors employed by the Employer' including 'statutory bodies'.

Sub-Clause (2) affords the Contractor the right of claim if compliance with Sub-Clause (1) above results 'in delay or cost beyond that reasonably foreseen by an experienced contractor at the time of tender'. Under such circumstances, the Engineer is entitled to certify payment of such extra cost and award an appropriate extension of time.

Sub-Clause (2) illustrates a very important principle in contracts i.e. what the Contractor can reasonably be expected to have allowed for in his rates. In this case, the Contractor must allow all reasonable facilities for other contractors. This does not mean that he must bear any costs which happen to be incurred. The only costs that the Contractor must bear are those which were included under the various Item Coverages in the method of measurement applicable to the contract at the time of tender, in this case the 'Method of Measurement for Highway Works', provided that details of the required facilities were included in the contract documents, usually in the drawings or in the specification. There has to be a specification which details what the Contractor must do in this respect. How can he reasonably be expected to have made allowance unless he knew at the time of tender what would be required? This important principle is considered further in relation to Item Coverages in the 'Method of Measurement for Highway Works' in Chapter 3.

This also demonstrates the very significant difference (particularly to Contractors) between responsibility and extra cost. The Contractor only bears the cost if the requirements were clear (i.e. were specified) at the time of tender and the rate within which such allowance should be included was clearly stated (i.e. in the appropriate Item Coverages) again at the time of tender. It is a constant source of surprise how many Engineers consider that the phrases employed in this Clause give the Contractor adequate information upon which to make some sort of financial allowance in his rates for providing unspecified facilities.

---

[Contractor's address]

[Engineer]

[Date]

Dear Sir

**[Contract description]**
**[Contract location]**
**[Location of delay]**
**Notice of delay by other contractor/authorised authority/statutory body**

The operations being carried out by [[your contractor/authorised authority/statutory body,] [           ]], have involved us in a delay of [           ] [days/weeks] and cost beyond that reasonably to be foreseen by an experienced contractor at the time of tender.

We hereby request that pursuant to Clause 31(2) of the Conditions of Contract, you grant an extension of time under the terms of Clause 44(1) of the Conditions of Contract and certify payment of this extra cost under the terms of Clause 60 of the Conditions of Contract.

We shall include a sum related hereto in the next monthly statement submitted in accordance with Clause 60(1) and confirm that appropriate contemporary records have been kept to support this application for an extension of time and additional payment. Please note that this letter constitutes a notice required pursuant to Clause 52(4)(b) of the Conditions of Contract.

Yours faithfully



B Ferry
Agent
for Unlimited Contracting Ltd


*Standard letter 12. Clause 31(2). Notice of claim as a result of delay and extra cost resulting from the operations of another contractor, authorised authority or statutory body*

---

## 2.13. Clause 36

### SUB-CLAUSE (1)—QUALITY OF MATERIALS AND WORKMANSHIP AND TESTS

All materials and workmanship shall be of the respective kinds described in the Contract and in accordance with the Engineer's instructions and shall be subjected from time to time to such tests as the Engineer may direct at the place of manufacture or fabrication or on the Site or such other place or places as may be specified in the Contract. The Contractor shall provide such assistance instruments machines labour and materials as are normally required for examining measuring and testing any work and the quality weight or quantity of any materials used and shall supply samples of materials before incorporation in the Works for testing as may be selected and required by the Engineer.

### SUB-CLAUSE (2)—COST OF SAMPLES

All samples shall be supplied by the Contractor at his own cost if the supply thereof is clearly intended by or provided for in the Contract but if not then at the cost of the Employer.

## SUB-CLAUSE (3)—COST OF TESTS

The cost of making any test shall be borne by the Contractor if such test is clearly intended by or provided for in the Contract and (in the cases only of a test under load or of a test to ascertain whether the design of any finished or partially finished work is appropriate for the purposes which it was intended to fulfil) is particularised in the Specification or Bill of Quantities in sufficient detail to enable the Contractor to have priced or allowed for the same in his Tender. If any test is ordered by the Engineer which is either

(a) not so intended by or provided for

(b) (in the cases above mentioned) is not so particularised

then the cost of such test shall be borne by the Contractor if the test shows the workmanship or materials not to be in accordance with the provisions of the Contract or the Engineer's instructions but otherwise by the Employer.

Sub-Clause (1) requires that all materials and workmanship are in accordance with the requirements of the Contract. It allows the Engineer to direct such tests as he thinks fit. Furthermore, it requires the Contractor to 'provide such assistance instruments machines labour and materials as are normally required for examining measuring and testing any work' and to supply samples as required by the Engineer.

Sub-Clauses (2) and (3) require the Contractor to supply samples (Sub-Clause (2)) and carry out tests at his own cost if the supply or execution thereof was clearly intended by or provided for in the contract but if not then at the cost of the Employer.

It is a common misconception that all sampling and testing is funded by the Contractor. Again, unless quantities of particular samples or tests are specified (they are listed in Appendix 1/5 in the contract as dictated by Volume 1: 'Specification for Highway Works') the Contractor cannot reasonably be expected to have allowed for them in his rates and will be entitled to recover the costs thereof. In a substantial number of cases the associated costs are insignificant but it is worthy of consideration when the numbers and frequency may warrant pursuit of costs. Contractors should also bear in mind that the cost of providing samples has two elements, the cost of the material itself and the cost of taking the sample.

[Contractor's address]

[Engineer]

[Date]

Dear Sir

**[Contract description]**
**[Contract location]**
**[Sampling/testing] of [            ]**
**Costs of [samples/testing]**

The amount of [specify material] has increased from [            ] to [            ] and, consequently, the costs associated with the provision of samples and testing are not allowed for in the Contract. We hereby give notice that we intend to seek recovery of all associated costs pursuant to Clause 36(2) and (3) of the Conditions of Contract.

Please note that this letter constitutes a notice required pursuant to Clause 52(4)(b) of the Conditions of Contract. Appropriate contemporary records will be kept to support an application for additional payment.

Yours faithfully


D Shannon
Agent
for Unlimited Contracting Ltd


*Standard letter 13. Clause 36(2)/(3). Notice of claim as a result of amount or types of sampling and testing rising above that suggested in the tender documents*

## 2.14. Clause 38

### SUB-CLAUSE (1)—EXAMINATION OF WORK BEFORE COVERING UP

No work shall be covered up or put out of view without the approval of the Engineer and the Contractor shall afford full opportunity for the Engineer to examine and measure any work which is about to be covered up or put out of view and to examine foundations before permanent work is placed thereon. The Contractor shall give due notice to the Engineer whenever any such work or foundations is or are ready or about to be ready for examination and the Engineer shall without unreasonable delay unless he considers it unnecessary and advises the Contractor accordingly attend for the purpose of examining and measuring such work or of examining such foundations.

### SUB-CLAUSE (2)—UNCOVERING AND MAKING OPENINGS

The Contractor shall uncover any part or parts of the Works or make openings in or through the same as the Engineer may from time to time direct and shall reinstate and make good such part or parts to the satisfaction of the Engineer. If any such part or parts have been covered up or put out of view after compliance with the requirements of sub-clause (1) of this Clause and are found to be executed in accordance with the Contract the cost of uncovering making openings in or through reinstating and making good the same shall be borne by the Employer but in any other case all such cost shall be borne by the Contractor.

Sub-Clause (1) requires the Contractor to give the Engineer (and therefore, usually, the Engineer's Representative) the opportunity to examine and measure any work including foundations before it is covered up or put out of view. The Contractor must give notice and, consequently, would be wise to establish some system with the agreement of the Engineer or Engineer's Representative which minimises any delays in getting the Engineer's approval. This is particularly important when operations involving a significant number of expensive pieces of plant are being undertaken such as large scale earthworks and surfacing operations.

Sub-Clause (2) empowers the Engineer to have any part or parts of the Works uncovered or openings made. If the Contractor has complied with Sub-Clause (1) and the work complies then the Employer pays; if the work does not comply, then the cost falls to the Contractor. When a proportion is found to comply then it is suggested that the payment to the Contractor of proportional costs is reasonable.

# THE 5TH EDITION OF THE ICE CONDITIONS OF CONTRACT | 59

Clause 39 deals with situations which arise including those under Clause 38 where the Contractor has not complied with the contractual requirements and has to make good giving the Engineer power to have non-compliant work removed etc.

---

[Contractor's address]

[Engineer]

[Date]

Dear Sir

**[Contract description]**
**[Contract location]**
**[Details of work uncovered]**
**Notice of delay and extra cost as a result of uncovering work found to comply with the Contract**

We refer to your instruction dated [           ] to uncover the [           ] issued pursuant to Clause 38(2) of the Conditions of Contract. Having previously complied with the requirements of this Clause and discovering that the work meets the requirements of the Contract, we have to advise you of our intention to recover the costs associated therewith in accordance with this Clause. Furthermore this instruction has delayed and disrupted the Works and we seek to recover the costs of this delay and have awarded an appropriate extension of time pursuant to Clause 13(3) of the Conditions of Contract.

This letter serves as a notice required pursuant to Clause 52(4)(b) of the Conditions of Contract. Appropriate contemporary records will be kept to support an application for additional payment and extension of time pursuant to Clause 44(1) of the Conditions of Contract.

Yours faithfully


P Skellern
Agent
for Unlimited Contracting Ltd

*Standard letter 14. Clause 38(2). Notice of delay and extra cost due to the uncovering of work found to be in compliance with the contract*

---

## 2.15. Clause 40

### SUB-CLAUSE (1)—SUSPENSION OF WORK

The Contractor shall on the written order of the Engineer suspend the progress of the Works or any part thereof for such time or times and in such manner as the Engineer may consider necessary and shall during such suspension properly protect and secure the work so far as is necessary in the opinion of the Engineer. Subject to Clause 52(4) the Contractor shall be paid in accordance with Clause 60 the extra cost (if any) incurred in giving effect to the Engineer's instructions under this Clause except to the extent that such suspension is

(a) otherwise provided for in the Contract

(b) necessary by reason of weather conditions or by some default on the part of the Contractor

(c) necessary for the proper execution of the work or for the safety of the Works or any part thereof inasmuch as such necessity does not arise from any act or default of the Engineer or the Employer or from any of the Excepted Risks defined in Clause 20.

The Engineer shall take any delay occasioned by a suspension order under this Clause (including that arising from any act or default of the Engineer or the Employer) into account in determining any extension of time to which the Contractor is entitled under Clause 44 except when such suspension is otherwise provided for in the Contract or is necessary by reason of some default on the part of the Contractor.

### SUB-CLAUSE (2)—SUSPENSION LASTING MORE THAN THREE MONTHS

If the progress of the Works or any part thereof is suspended on the written order of the Engineer and if permission to resume work is not given by the Engineer within a period of three months from the date of suspension then the Contractor may unless such suspension is otherwise provided for in the Contract or continues to be necessary by reason of some default on the part of the Contractor serve a written notice on the Engineer requiring permission within 28 days from the receipt of such notice to proceed with the Works or that part thereof in regard to which progress is suspended. If within the 28 days the Engineer does not grant such permission the Contractor by a further written notice so served may (but is not bound to) elect to treat the suspension where it affects part only of the Works as an omission of such part under Clause 51 or where it affects the whole Works as an abandonment of the Contract by the Employer.

Sub-Clause (1) permits the Engineer to order the suspension of the Works or part thereof in writing in such manner as he prescribes. The Contractor has to protect and secure the Works so far as is necessary in the opinion of the Engineer. The Employer will pay any extra cost incurred by the Contractor where the Works or part thereof are suspended unless the suspension is

(*a*) specified in the contract
(*b*) due to weather conditions or default on the part of the Contractor
(*c*) necessary for the proper execution or safety of the Works or part thereof provided that the suspension is not due to any act or default of the Engineer (e.g. failure to provide drawings) or Employer (e.g. failure to give possession of all or part of the site) or from any of the excepted risks (riot, war, invasion, act of foreign enemies, hostilities, civil war, rebellion, revolution, insurrection etc.—see Clause 20(3)).

Where the suspension is due only in part to any of the circumstances set out above then the Contractor is entitled to extra cost for that proportion which is not covered by the above.

The Engineer is required to grant an extension of time for the period of the suspension unless the suspension is specified in the contract or due to default by the Contractor.

Sub-Clause (2) covers the situation where suspension extends for a period longer than three months. Providing the suspension is not for any of the reasons set out in (*a*), (*b*) or (*c*) above, then the Contractor can, in writing, request permission to proceed with the Works or the part of the Works which is the subject of the suspension. If within 28 days of the receipt of such notice the Engineer does not grant permission then the Contractor can

(i) treat the contract as abandoned if the whole of the Works is affected
(ii) treat the affected part as if it was cancelled i.e. an omission under Clause 51.

[Contractor's address]

[Engineer]

[Date]

Dear Sir

**[Contract description]**
**[Contract location]**
**[Specify extent of suspension]**
**Notice of delay and additional cost as a result of a suspension**

We refer to your letter dated [            ] ordering a suspension of the Works and wish to record that the suspension is not occasioned by any of the circumstances listed in exclusions (a), (b) or (c) of Clause 40(1) of the Conditions of Contract. Furthermore, giving effect to this instruction has incurred us in extra cost and caused a delay of [            ] [days/weeks]. We hereby request that pursuant to the Clause 40(1) of the Conditions of Contract, you grant an extension of time under the terms of Clause 44(1) of the Conditions of Contract and certify payment of our extra cost under the terms of Clause 60 of the Conditions of Contract.

We shall include a sum related hereto in the next monthly statement submitted in accordance with Clause 60(1) and confirm that appropriate contemporary records have been kept to support this application for an extension of time and additional payment. Please note that this letter constitutes a notice required pursuant to Clause 52(4)(b) of the Conditions of Contract.

Yours faithfully



R Johnson
Agent
for Unlimited Contracting Ltd


*Standard letter 15. Clause 40(1). Notice of claim as a result of suspension of the Works or part thereof*

## 2.16. Clause 41

### COMMENCEMENT OF WORKS

The Contractor shall commence the Works on or as soon as is reasonably possible after the Date for Commencement of the Works to be notified by the Engineer in writing which date shall be within a reasonable time after the date of acceptance of the Tender. Thereafter the Contractor shall proceed with the Works with due expedition and without delay in accordance with the Contract.

The Engineer should write to the Contractor advising of the Date for Commencement of the Works. The Engineer would normally discuss this matter with the Contractor before issuing a notice. The Contractor may be able to influence the Engineer by suggesting a date.

The Date for Commencement of the Works starts the clock as far as completion and, therefore, liability for the payment of liquidated damages (as discussed later in relation to Clause 47) is concerned.

## 2.17. Clause 42

### SUB-CLAUSE (1)—POSSESSION OF SITE

Save in so far as the Contract may prescribe the extent of portions of the Site of which the Contractor is to be given possession from time to time and the order in which such portions shall be made available to him and subject to any requirement in the Contract as to the order in which the Works shall be executed the Employer will at the Date for Commencement of the Works notified under Clause 41 give to the Contractor possession of so much of the Site as may be required to enable the Contractor to commence and proceed with the construction of the Works in accordance with the programme referred to in Clause 14 and will from time to time as the Works proceed give to the Contractor possession of such further portions of the Site as may be required to enable the Contractor to proceed with the construction of the Works with due dispatch in accordance with the said programme. If the Contractor suffers delay or incurs cost from failure on the part of the Employer to give possession in accordance with the terms of this Clause then the Engineer shall take such delay into account in determining any extension of time to which the Contractor is entitled under Clause 44 and the Contractor shall subject to Clause 52(4) be paid in accordance with Clause 60 the amount of such cost as may be reasonable.

## SUB-CLAUSE (2)—WAYLEAVES, ETC

The Contractor shall bear all expenses and charges for special or temporary wayleaves required by him in connection with access to the Site. The Contractor shall also provide at his own cost any additional accommodation outside the Site required by him for the purposes of the Works.

Under Sub-Clause (1), the Employer may give possession of the site under a variety of options. The availability and nature of accesses may be specified and the order in which the Works are to be constructed can be directed all as specified in the contract. Subject to these contractual provisions, the Employer is required to give the Contractor possession of the site to the extent necessary to carry out construction of the Works in accordance with the Clause 14 programme and at the Date for Commencement of the Works (notified by the Engineer to the Contractor pursuant to Clause 41). Where the Employer fails to do so then the Engineer has the right to grant an extension of time pursuant to Clause 44, and, as usual, subject to the Contractor's rights under Clause 52(4), certify payment in accordance with Clause 60(2) ('Monthly Payments').

This Clause would be enacted if possession was not given as suggested in the contract or in order to meet the programme. It would also be invoked if the Employer's other contractors, including utilities, occupied part of the site which was supposed to be available and such occupation impeded progress.

---

[Contractor's address]

[Engineer]

[Date]

Dear Sir

**[Contract description]**
**[Contract location]**
**[Specify extent of non-possession]**
**Notice of delay and additional cost as a result of late possession**

We write to record that we were not given possession of the above at Date for Commencement of the Works notified to us by you pursuant to Clause 41 of the Conditions of Contract in your letter dated [            ]. As a result, we have suffered a delay of [            ] [days/weeks] and incurred cost.

We hereby request that pursuant to Clause 42(1) of the Conditions of Contract, you grant an extension of time under the terms of Clause 44(1) of the Conditions of Contract and certify payment of this extra cost under the terms of Clause 60 of the Conditions of Contract.

We shall include a sum related hereto in the next monthly statement submitted in accordance with Clause 60(1) and confirm that appropriate contemporary records have been kept to support this application for an extension of time and additional payment. Please note that this letter constitutes a notice required pursuant to Clause 52(4)(b) of the Conditions of Contract.

Yours faithfully


B B Broonzy
Agent
for Unlimited Contracting Ltd


*Standard letter 16. Clause 42(1). Notice of claim as a result of non-possession of all or part of the site*

---

[Contractor's address]

[Engineer]

[Date]

Dear Sir

**[Contract description]**
**[Contract location]**
**[Specify extent of non-possession]**
**Notice of delay and additional cost as a result of limited possession**

Further to our letter to you dated [            ], we write to record that due to the late completion of [            ] by [another contractor], we were not given possession of the above until [            ]. This is [            ] [days/weeks] later than that specified in the Contract. As a result, we have suffered delay and incurred cost.

We hereby request that pursuant to Clause 42(1) of the Conditions of Contract, you grant an extension of time under the terms of Clause 44(1) of the Conditions of Contract and certify payment of this extra cost under the terms of Clause 60 of the Conditions of Contract.

We shall include a sum related hereto in the next monthly statement submitted in accordance with Clause 60(1) and confirm that appropriate contemporary records have been kept to support this application for an extension of time and additional payment. Please note that this letter constitutes a notice required pursuant to Clause 52(4)(b) of the Conditions of Contract.

Yours faithfully


P A Green
Agent
for Unlimited Contracting Ltd


*Standard letter 17. Clause 42(1). Notice of claim as a result of late possession of part of the site due to operations by another contractor*

## 2.18. Clause 43

**TIME FOR COMPLETION**

> The whole of the Works and any Section required to be completed within a particular time as stated in the Appendix to the Form of Tender shall be completed within the time so stated (or such extended time as may be allowed under Clause 44) calculated from the Date for Commencement of the Works notified under Clause 41.

The Contractor has to finish within the time stated in the Appendix to the Form of Tender. Note the use of the word 'within': in other words he can finish early without penalty. As discussed earlier it should be remembered that there is no implied term requiring the Employer or his agents (including the Engineer) to perform their own obligations earlier in order to facilitate early completion by the Contractor (*Glenlion Construction v. Guinness Trust* (1988) 39 BLR 89). Under Clause 7(3), the Contractor has a right to extension of time and reasonable cost if delay or cost is incurred as a result of the Engineer failing to issue drawings or instructions necessary for the purpose of the 'adequate construction completion and maintenance of the Works' (Clause 7(1)).

The danger for the Contractor is that little consideration has actually been given to the time for completion by the Engineer; in practice this is often the case. No calculation is deemed necessary: '26 weeks should be enough' is often as scientific as the process becomes. The Contractor starts the job, works reasonably diligently, and is awarded the occasional extension of time but ends up paying liquidated damages to the Employer. What can he do? Nothing. Could a plea have been made for extended time during the currency of the Works? No, he has signed a contract to deliver the Works within the period stated. The parties have agreed to that express term. All Engineers should give careful thought to what is a reasonable period for completing the Works. During the tender stage, Contractors should ensure that the time for completion is achievable and if not then make appropriate allowance in financial or executional terms.

## 2.19. Clause 44

### SUB-CLAUSE (1)—EXTENSION OF TIME FOR COMPLETION

Should any variation ordered under Clause 51(1) or increased quantities referred to in Clause 51(3) or any other cause of delay referred to in these Conditions or exceptional adverse weather conditions or other special circumstances of any kind whatsoever which may occur be such as fairly to entitle the Contractor to an extension of time for the completion of the Works (where different periods for completion of different Sections are provided for in the Appendix to the Form of Tender) of the relevant Section the Contractor shall within 28 days after the cause of any delay has arisen or as soon thereafter as is reasonable in all the circumstances deliver to the Engineer full and detailed particulars of any claim to extension of time to which he considers himself entitled in order that such claim may be investigated at the time.

## SUB-CLAUSE (2)—INTERIM ASSESSMENT OF EXTENSION

The Engineer shall upon receipt of such particulars or if he thinks fit in the absence of any such claim consider all the circumstances known to him at that time and make an assessment of the extension of time (if any) to which he considers himself entitled for the completion of the Works or relevant Section and shall by notice in writing to the Contractor grant such extension of time for completion. In the event that the Contractor shall have made a claim for an extension of time but the Engineer considers the Contractor not entitled thereto the Engineer shall so inform the Contractor.

## SUB-CLAUSE (3)—ASSESSMENT AT DUE DATE FOR COMPLETION

The Engineer shall at or as soon as possible after the due date or extended date for completion (and whether or not the Contractor shall have made any claim for an extension of time) consider all the circumstances known to him at that time and take action similar to that provided for in sub-clause (2) of this Clause. Should the Engineer consider that the Contractor is not entitled to an extension of time he shall so notify the Employer and the Contractor.

## SUB-CLAUSE (4)—FINAL DETERMINATION OF EXTENSION

The Engineer shall upon the issue of the Certificate of Completion of the Works or of the relevant Section review all the circumstances of the kind referred to in sub-clause (1) of this Clause and shall finally determine and certify to the Contractor with a copy to the Employer the overall extension of time (if any) to which he considers the Contractor entitled in respect of the Works or the relevant Section. No such final review of the circumstances shall result in a decrease in any extension of time already granted by the Engineer pursuant to sub-clauses (2) or (3) of this Clause.

Sub-Clause (1) entitles the Contractor to an extension of time for

(a) any variation under Clause 51(1) ('Ordered Variations')
(b) increased quantities under Clause 51(3) ('Changes in Quantities')
(c) any other cause of delay specified in the *5th Edition of the ICE Conditions of Contract*
(d) exceptional adverse weather conditions or
(e) other special circumstances.

Within 28 days or such time as is reasonable thereafter the Contractor must deliver to the Engineer full details of the claim to extension of time which is being sought.

Sub-Clause (2) requires the Engineer to consider, on receipt of the details cited in Sub-Clause (1) above, all the circumstances known to him and grant any due extension in writing. It also allows the Engineer to make his own assessment where no application has been made by the Contractor. If the Engineer decides to reject the Contractor's request for an extension of time he is required to advise the Contractor but not necessarily in writing. If this should be the case then, it is suggested, the Contractor may be wise to supply written confirmation.

Under Sub-Clause (3), at the Date for Completion (or extended date if an extension has already been granted) the Engineer must repeat the process outlined in Sub-Clause (2) and consider all the information known to him including requests by the Contractor for extensions of time and grant the appropriate extension.

Under Sub-Clause (4), upon the issue of Certificate of Completion for the Works, the Engineer must make a final decision on any extensions of time. Note that the Engineer cannot, at this stage, rescind any extension or extensions of time which have previously been granted.

This is a most important Clause. Claims often originate as a simple request for an extension of time. Many claims include a request for an extension of time as well as costs. It is vital that the Contractor applies for extensions of time for all incidents which so warrant. Substantial sums are often associated with the granting of extensions of time.

Note the provision for the Engineer to grant an extension of time for 'other special circumstances of any kind whatsoever'. The Contractor should bear this in mind in the light of events. For example, there is no contractual right to extension of time if extra testing is ordered (see Clause 36) but if this event delayed the execution of the Works then this term may be useful.

The Engineer may fail to exercise his obligations under this Clause and, in such circumstances, the Contractor should prompt the Engineer to do so.

If the Contractor makes an application for an extension of time which is rejected (whether or not it is subsequently granted) and the Contractor takes steps to make up the time (acceleration) in order to avoid liability for the payment of liquidated damages then he may have a claim for breach of contract on the grounds that the Employer's agent, namely the Engineer, has failed to exercise his powers properly under the terms of Clause 44(1).[7]

There are three occasions when the Engineer is obliged to review the need to grant extensions of time

(a) upon receipt of particulars (Clause 44(2))
(b) at or as soon as possible after the due date or (if already the subject of an extension of time) the extended date for completion and
(c) upon the issue of the Certificate of Completion for the whole of the Works or a section (Clause 48).

Note that the review under (c) above cannot reduce any extension previously awarded under (a) or (b).

Note that no costs can be paid to the Contractor when an extension of time is due to exceptionally adverse weather conditions. What constitutes exceptional is open to debate, but the occurrence of snow in northern Scotland in the middle of January is not exceptional whereas the same weather conditions in south-west England may well be so. A Contractor may find it worthwhile to obtain copies of appropriate weather records for comparison but should always send copies of the weekly 'time lost' record to the Engineer so that there is no dispute in the compilation of final figures of time lost at the end of the contract.

The lack of a contractual mechanism for granting extensions of time under various circumstances even where there is no fault would prevent the Employer from applying liquidated damages[8] pursuant to Clause 47. Engineers should bear that fact in mind. Nowadays, Engineers and Employers have learned how to apply liquidated damages provisions properly and these can amount to very considerable sums of money. As a consequence, and in order to reduce or eliminate the deduction of liquidated damages, Contractors will produce claims for extensions of time. In order to minimise the possibility of liquidated damages being applied, Contractors should prompt the Engineer to grant or apply for, as appropriate, extensions of time whenever the circumstances so warrant.

[Contractor's address]

[Engineer]

[Date]

Dear Sir

**[Contract description]**
**[Contract location]**
**[Specify event which warrants extension]**
**Application for extension of time**

We refer to [the variation ordered pursuant to Clause 51(1) or (2)/ increased quantities pursuant to Clause 51(3)/other special circumstances] and confirm that this has caused a delay of [          ] [days/weeks]. Pursuant to Clause 44(1) of the Conditions of Contract we attach a schedule showing full and detailed particulars thereof.

We request that you grant an interim assessment of extension of time pursuant to Clause 44(2) of the Conditions of Contract.

Yours faithfully


B B King
Agent
for Unlimited Contracting Ltd

Enc


*Standard letter 18. Clause 44(1)/(2). Application for extension of time*

---

## 2.20.  Clause 47

### SUB-CLAUSE (1)—LIQUIDATED DAMAGES FOR WHOLE OF WORKS

(a) In the Appendix to the Form of Tender under the heading 'Liquidated Damages for Delay' there is stated in column 1 the sum which represents the Employer's genuine pre-estimate (expressed as a rate per week or per day as the case may be) of the damages likely to be suffered by him in the event that the whole of the Works shall not be completed within the time prescribed by Clause 43.

Provided that in lieu of such sum there may be stated such lesser sum as represents the limit of the Contractor's liability for damages for failure to complete the whole of the Works within the time for completion therefor or any extension thereof granted under Clause 44.

(b) If the Contractor should fail to complete the whole of the Works within the prescribed time of any extension thereof granted under Clause 44 the Contractor shall pay to the Employer for such default the sum stated in column 1 aforesaid for every week or day as the case may be which shall elapse between the date on which the prescribed time or any extension thereof expired and the date of completion of the whole of the Works. Provided that if any part of the Works not being a Section or part of a Section shall be certified as complete pursuant to Clause 48 before completion of the whole of the Works the sum stated in column 1 shall be reduced by the proportion which the value of the part completed bears to the value of the whole of the Works.

## SUB-CLAUSE (2)—LIQUIDATED DAMAGES FOR SECTIONS

(a) In cases where any Section shall be required to be completed within a particular time as stated in the Appendix to the Form of Tender there shall also be stated in the said Appendix under the heading 'Liquidated Damages for Delay' in column 2 the sum by which the damages stated in column 1 or the limit of the Contractor's said liability as the case may be shall be reduced upon completion of each such Section and in column 3 the sum which represents the Employer's genuine pre-estimate (expressed as aforesaid) of any specific damage likely to be suffered by him in the event that such Section shall not be completed within that time.

Provided that there may be stated in column 3 in lieu of such sum such lesser sum as represents the limit of the Contractor's liability for failure to complete the relevant Section within the relevant time.

(b) If the Contractor should fail to complete any Section within the relevant time for completion or any extension thereof granted under Clause 44 the Contractor shall pay to the Employer for such default the sum stated in column 3 aforesaid for every week or day as the case may be which shall elapse between the date on which the

# THE 5TH EDITION OF THE ICE CONDITIONS OF CONTRACT | 73

relevant time or any extension thereof expired and the date of completion of the relevant Section. Provided that:

(i) if completion of a Section shall be delayed beyond the due date for completion of the whole of the Works the damages payable under sub-clauses (1) and (2) of this Clause until completion of that Section shall be the sum stated in column 1 plus in respect of that Section the sum stated in column 3 less the sum stated in column 2

(ii) if any part of a Section shall be certified as complete pursuant to Clause 48 before completion of the whole thereof the sums stated in columns 2 and 3 in respect of that Section shall be reduced by the proportion which the value of the part bears to the value of the Section and the sum stated in column 1 shall be reduced by the same amount as the sum in column 2 is reduced

(iii) upon completion of any such Section the sum stated in column 1 shall be reduced by the sum stated in column 2 in respect of that Section at the date of such completion.

## SUB-CLAUSE (3)—DAMAGES NOT A PENALTY

All sums payable by the Contractor to the Employer pursuant to this Clause shall be paid as liquidated damages for delay and not as a penalty.

## SUB-CLAUSE (4)—DEDUCTION OF LIQUIDATED DAMAGES

If the Engineer shall under Clause 44(3) or (4) have determined and certified any extension of time to which he considers the Contractor entitled or shall have notified the Employer and the Contractor that he is of the opinion that the Contractor is not entitled to any or any further extension of time the Employer may deduct and retain from any sum otherwise payable by the Employer to the Contractor hereunder the amount which in the event that the Engineer's said opinion should not be subsequently revised would be the amount of the liquidated damages payable by the Contractor under this Clause.

## SUB-CLAUSE (5)—REIMBURSEMENT OF LIQUIDATED DAMAGES

If upon a subsequent or final review of the circumstances causing delay the Engineer shall grant an extension or further extension of time or if an arbitrator appointed under Clause 66 shall decide that the Engineer should have granted such an extension or further extension of time the Employer shall no longer be entitled to liquidated damages in respect of the period of such extension of time. Any sums in respect of such period which may have been recovered pursuant to sub-clause (3) of this Clause shall be reimbursable forthwith to the Contractor together with interest at the rate provided for in Clause 60(6) from the date on which such liquidated damages were recovered from the Contractor.

Sub-Clause (1) entitles the Employer to deduct a sum of money as a measure of loss where the contract has not reached substantial completion (technically the issue of a Certificate of Completion—see Clause 48) by the date for completion. This sum has to be a genuine pre-estimate of loss (*Clydebank Engineering and Shipbuilding Co v. Don Jose Ramos Yzquierdo-y-Castaneda* (1905) AC6).

Sub-Clause (2) covers the situation where the Works have been divided into sections in the original contract or the Contractor has persuaded the Engineer to issue a Certificate of Completion for a part of the Works. The Contractor should always apply for sectional completion where the circumstances so warrant e.g. a roundabout, a structure etc. Indeed, it may be that one of the factors which influences the order of procedure within the Contractor's programme is the prospect of achieving partial completion. Unless it is specified in the Contract, the Employer has no right to use part of the Works until the whole of the Works has been certified as substantially complete. Indeed, should the Employer wish to use part of the whole then the Contractor would be foolish to permit this without the issue of an appropriate Certificate of Completion for that part of the Works (but see also Clause 48(2)). The effect of such partial completion is a reduction in the amount due as liquidated damages for the whole of the Works. This Clause does not grapple with that particular issue and it would be advisable for the Contractor to supply a calculation which maximises the reduction by choosing the appropriate parameter e.g. proportion of total length of road or proportion of total cost of the Works etc. Other desirable results of achieving partial completion, as far as the Contractor is concerned, are the commencement of the defects liability period, the reduction in insurance liability and the payment of monies withheld as retention pursuant to Clause 60(4).

Sub-Clause (3) confirms that the sums payable for damages are not a penalty although actually this is unnecessary because the courts have long since established that principle (*Clydebank*

# THE 5TH EDITION OF THE ICE CONDITIONS OF CONTRACT | 75

*Engineering and Shipbuilding Co v. Don Jose Ramos Yzquierdo-y-Castaneda* (1905) AC6).

Sub-Clause (4) requires the Engineer to have carried out the assessment processes for extensions of time specified in Clause 44(3) or (4) before the Employer is entitled to liquidated damages. This Sub-Clause also empowers the Employer to deduct the sum from monies due to the Contractor. If the Clause 44(3) or (4) assessment has not been carried out then the deduction of liquidated damages is not permissible. This process must be carried out in full including the Engineer notifying the Contractor that he considers that no extension of time is due.

Sub-Clause (5) recognises the situation where an extension of time is granted subsequently either by the Engineer or by an arbitrator and liquidated damages have been deducted. These are then repaid along with interest as per late payment on monthly statements as per Clause 60.

Liquidated damages are seen by the courts as a fair way of paying a measure of damages to one party if the other party is in breach in terms of a provision for completion. Both parties know the amount payable in advance and so they know what the expenditure or income, as appropriate, will be in the event of breach. It may be that the actual loss is very much more or less than this amount; that is irrelevant. The amount payable cannot be increased or reduced. The Contractor may be wise to pay liquidated damages, because if he manages to have them set aside, he may face a claim for damages through the courts, resulting in his having to pay more.

Note that where the sum for liquidated damages is stated as weeks the Employer cannot deduct part of a week's liquidated damages.

The origin of many requests for extension of time is the reduction or elimination of liquidated damages which have been applied or are expected to be applied.

Engineers have, after many years, finally got to grips with their obligations under this Clause and now religiously carry out what they deem to be the 'genuine pre-estimate of loss' process. Contractors should be extremely aware of the possible consequences of liquidated damages provisions and should assess the effects at the tender stage. Engineers should be equally conscious of the effects of variations, weather conditions or other matters which cause delay and disruption and assess fairly on the basis of prevailing events.

(For more on this subject refer to Chapter 3 on the section relating to the 'Model Contract Documents', Volume 0, Section 1, Parts 2 to 5 of the *Manual of Contract Documents for Highway Works*.)

## 2.21. Clause 48

### SUB-CLAUSE (1)—CERTIFICATE OF COMPLETION OF WORKS

When the Contractor shall consider that the whole of the Works has been substantially completed and has satisfactorily passed any final test that may be prescribed by the Contract he may give a notice to that effect to the Engineer or to the Engineer's Representative accompanied by an undertaking to finish any outstanding work during the Period of Maintenance. Such notice and undertaking shall be in writing and shall be deemed to be a request by the Contractor for the Engineer to issue a Certificate of Completion in respect of the Works and the Engineer shall within 21 days of the date of delivery of such notice either issue to the Contractor (with a copy to the Employer) a Certificate of Completion stating the date on which in his opinion the Works were substantially completed in accordance with the Contract or else give instructions in writing to the Contractor specifying all the work which in the Engineer's opinion requires to be done by the Contractor before the issue of such certificate. If the Engineer shall give such instructions the Contractor shall be entitled to receive such Certificate of Completion within 21 days of completion to the satisfaction of the Engineer of the work specified by the said instructions.

### SUB-CLAUSE (2)—COMPLETION OF SECTIONS AND OCCUPIED PARTS

Similarly in accordance with the procedure set out in sub-clause (1) of this Clause the Contractor may request and the Engineer shall issue a Certificate of Completion in respect of

(a) any Section in respect of which a separate time for completion is provided in the Appendix to the Form of Tender

(b) any substantial part of the Works which has been both completed to the satisfaction of the Engineer and occupied or used by the Employer.

### SUB-CLAUSE (3)—COMPLETION OF OTHER PARTS OF WORKS

If the Engineer shall be of the opinion that any part of the Works shall have been substantially completed and shall have satisfactorily passed any final test that may be prescribed by the Contract he may issue a Certificate of Completion in respect of that part of the Works before completion of the whole of the Works and upon the issue of such certificate the Contractor shall be deemed to have undertaken to complete any outstanding work in that part of the Works during the Period of Maintenance.

## SUB-CLAUSE (4)—REINSTATEMENT OF GROUND

Provided always that a Certificate of Completion given in respect of any Section or part of the Works before completion of the whole shall not be deemed to certify completion of any ground or surfaces requiring reinstatement unless such certificate shall expressly so state.

Sub-Clause (1) entitles the Contractor to request the Engineer to issue a Certificate of Completion for the whole of the Works when the Contractor considers that the whole is substantially complete. Within 21 days the Engineer must respond by either granting the request stating the date upon which the Works were substantially complete or else give instruction to the Contractor indicating those items which must be executed before a Certificate is granted. What constitutes 'substantial' is not specified but the use of the phrase 'accompanied by an undertaking to finish any outstanding works during the Period of Maintenance' indicates that completion does not constitute every item of work which is specified in the contract. The 'period of maintenance' is defined in Clause 49(1) and is stated in the Appendix to the Form of Tender. It might seem fair to interpret the word 'substantial' as being related to the function of the finished Works as a whole so, for example, in a roadworks contract the lack of some soiling and seeding would not appear to hinder the use of the Works for their intended purpose. It is up to the Contractor to make the best possible case in the prevailing circumstances.

Sub-Clause (2) repeats the procedure set out in Sub-Clause (1) for sections which are specified in the contract and a 'substantial part' of the Works which is acceptable to the Engineer and occupied or used by the Employer. Again the word 'substantial' appears and the approach to be used in interpreting this word in the previous paragraph seems reasonable where circumstances permit.

Sub-Clause (3) confers upon the Engineer a discretionary power to issue a Certificate of Completion for a section which he considers is substantially completed and has passed any final testing.

As has been emphasised, the granting of a Certificate of Completion releases the Contractor from all or part of the liquidated damages provisions. However, it also has other important ramifications, financial and otherwise: the commencement of the defects liability period, the reduction in insurance liability and the payment of monies withheld as retention pursuant to Clause 60(4). Contractors must make application whenever the circumstances warrant.

Note that work done in the maintenance period under the standard undertaking does not carry any extension of maintenance period.

Note also that the Engineer is not obliged to grant a Certificate of Completion until the Contractor makes application except as set out under Sub-Clause (3). It is difficult to be sympathetic to any Contractor who fails to request formal completion.

---

[Contractor's address]

[Engineer]

[Date]

Dear Sir

**[Contract description]**
**[Contract location]**
**Application for Certificate of Completion for the whole of the Works**

We consider that the whole of the Works has been substantially completed on [          ] and hereby give notice to that effect pursuant to Clause 48(1) of the Conditions of Contract. We undertake to finish any outstanding work during the period of maintenance. Accordingly, we request that you issue a Certificate of Completion in respect of the whole of the Works.

Yours faithfully


J L Hooker
Agent
for Unlimited Contracting Ltd


*Standard letter 19. Clause 48(1). Application for Certificate of Completion for the whole of the Works*

[Contractor's address]

[Engineer]

[Date]

Dear Sir

**[Contract description]**
**[Contract location]**
**[Specify section]**
**Application for Certificate of Completion for a section of the Works**

We consider that the above section has been substantially completed on [              ] and hereby give notice to that effect pursuant to Clause 48(2)(a) of the Conditions of Contract. We undertake to finish any outstanding work during the period of maintenance. Accordingly, we request that you issue a Certificate of Completion in respect of this section of the Works.

Yours faithfully


O Redding
Agent
for Unlimited Contracting Ltd


*Standard letter 20. Clause 48(2)(a). Application for Certificate of Completion for a section of the Works*

[Contractor's address]

[Engineer]

[Date]

Dear Sir

**[Contract description]**
**[Contract location]**
**[Specify section]**
**Application for Certificate of Completion for part of the Works**

We consider that the above section constitutes a substantial part of the Works. It was substantially completed on [          ] and we hereby give notice to that effect pursuant to Clause 48(2)(b) of the Conditions of Contract. We undertake to finish any outstanding work during the period of maintenance. Accordingly, we request that you issue a Certificate of Completion in respect of this section of the Works.

Yours faithfully


I Hayes
Agent
for Unlimited Contracting Ltd


*Standard letter 21. Clauses 48(2)(b). Application for Certificate of Completion for part of the Works*

---

In standard letters 19, 20 and 21, the date when the Contractor considers that the Works or part thereof were completed is included. This is designed to prompt the Engineer to issue the Certificate of Completion as of that specified date rather than the date when he issues the Certificate.

## 2.22. Clause 50

### CONTRACTOR TO SEARCH

The Contractor shall if required by the Engineer in writing carry out such searches tests or trials as may be necessary to determine the cause of any defect imperfection or fault under the directions of the Engineer. Unless such defect imperfection or fault shall be one for which the Contractor is liable under the Contract the cost of the work carried out by the Contractor as aforesaid shall be borne by the Employer. But if such defect imperfection or fault shall be one for which the Contractor is liable the cost of the work carried out as aforesaid shall be borne by the Contractor and he shall in such case repair rectify and make good such defect imperfection or fault at his own expense in accordance with Clause 49.

This entitles the Engineer to instruct the Contractor to carry out searches, tests or trials to determine the cause of any defect etc. Unless the latter is found to be the fault of the Contractor, in which case the costs of reparation fall upon the Contractor in accordance with Clause 49 ('Cost of Execution of Work of Repair, etc'), then the cost is met by the Employer. There is no mention made of apportionment where a proportion of the items are checked but it is suggested that it would seem reasonable for the Contractor to expect to be paid for that segment which is acceptable.

The reference in the last line of this Clause implies that it can only be invoked during the period of maintenance which is covered in Clause 49.

---

[Contractor's address]

[Engineer]

[Date]

Dear Sir

**[Contract description]**
**[Contract location]**
**[Specify which searches tests or trials]**
**Cost of [searches/tests/trials]**

We hereby apply for reimbursement for the costs of the above which showed that the [           ] complies with the requirements of the contract.

We hereby request that pursuant to Clause 50 of the Conditions of Contract, you certify payment of these costs under the terms of Clause 60 of the Conditions of Contract.

We shall include a sum related hereto in the next monthly statement submitted in accordance with Clause 60(1) and confirm that appropriate contemporary records have been kept to support this application for an extension of time and additional payment. Please note that this letter constitutes a notice required pursuant to Clause 52(4)(b) of the Conditions of Contract.

Yours faithfully


C Burnett
Agent
for Unlimited Contracting Ltd


*Standard letter 22. Clause 50. Contractor's request for payment for searches, tests or trials*

## 2.23. Clause 51

### SUB-CLAUSE (1)—ORDERED VARIATIONS

The Engineer shall order any variation to any part of the Works that may in his opinion be necessary for the completion of the Works and shall have the power to order any variation that for any other reason shall in his opinion be desirable for the satisfactory completion and functioning of the Works. Such variations may include additions omissions substitutions alterations changes in quality form character kind position dimension level or line and changes in the specified sequence method or timing of construction (if any).

### SUB-CLAUSE (2)—ORDERED VARIATIONS TO BE IN WRITING

No such variation shall be made by the Contractor without an order by the Engineer. All such orders shall be given in writing provided that if for any reason the Engineer shall find it necessary to give any such order orally in the first instance the Contractor shall comply with such oral order. Such oral order shall be confirmed in writing by the Engineer as soon as is possible in the circumstances. If the Contractor shall confirm in writing to the Engineer any oral order by the Engineer and such confirmation shall not be contradicted in writing by the Engineer forthwith it shall be deemed to be an order in writing by the Engineer. No variation ordered or deemed to be ordered in writing in accordance with sub-clauses (1) and (2) of this Clause shall in any way vitiate or invalidate the Contract but the value (if any) of all such variations shall be taken into account in ascertaining the amount of the Contract Price.

## SUB-CLAUSE (3)—CHANGES IN QUANTITIES

No order in writing shall be required for increase or decrease in the quantity of any work where such increase or decrease is not the result of an order given under this Clause but is the result of the quantities exceeding or being less than those stated in the Bill of Quantities.

Sub-Clause (1) allows the Engineer to order any variation to any part of the Works. Variations include 'additions omissions substitutions alterations changes in quality form character kind position dimension level or line and changes in any specified sequence method or timing of construction'. Note the extent of events which constitute a variation providing that the event is 'necessary' or 'desirable'. Where the change offers advantage only to the Contractor but conveys no benefit to the Employer then it does not require an ordered variation. The Engineer may however agree to the change subject to his terms. In the normal course of construction events, it is difficult to find a change which does not constitute a variation assuming it is necessary or desirable but see Sub-Clause (3) related to increased or decreased quantities. However, all the types of variations which are permitted here are within the context of the original contract and, consequently, the Engineer does not have the right to order the Contractor, for example, to extend the area of the contract. An example would be where a contract is being carried out in an urban area for surface dressing in several streets specified in the contract. The Engineer would have some difficulty in adding another street to those in the contract and expecting those to be done at billed rates. Equally, he would not be able to order the Contractor to execute additional work involving a different process say, continuing the earlier example, instructing the use of a modified binder where there is no similar material in the original contract. In such circumstances, the Contractor has the right to refuse to carry out the work or may do so at prices which are outwith the scope of the pricing regime adopted in the original contract.

The effects of a number of individually insignificant variations can be appreciable in combination. Furthermore, each variation involves the Contractor in costs encountered simply administering it.

Sub-Clause (2) requires all variations to be in writing but permits the Engineer to issue oral instructions if necessary. What constitutes necessity is not defined. If the Contractor carries out work which constitutes a variation to the contract and has not been ordered by the Engineer then he is in breach of contract and is not entitled to payment for this work. Ironically this would include changes which are suggested by the Contractor. Often, changes in the specified materials are agreed with the Engineer but both the Engineer and Contractor should, providing it is necessary or desirable, treat it as an ordered variation under this Clause and it should be subject to the normal notice requirements. If it is not the subject of written confirmation either by Engineer or Contractor then the Contractor is in a very difficult position if the Engineer subsequently denies agreement. This means that technically the varied work constitutes a breach of contract. Engineers are often reluctant to order a variation which results from a suggestion by the Contractor but the fact is that (normally) the contract cannot be varied without such an ordered variation. The reason behind the reluctance is that extra costs are often associated with ordered variations. If this is the underlying reason then the Engineer should seek to reach agreement with the Contractor before agreeing to the change unless there are economic or other advantages which are overwhelmingly compelling.

Sub-Clause (2) covers the situation where the Engineer chooses to issue oral variations; the Contractor is strongly advised to confirm the variation in writing to the Engineer at the earliest possible time and certainly before the varied work is carried out. If such confirmation is not contradicted in writing forthwith it is deemed to be an order in writing by the Engineer. Sub-Clause (2) goes on to state that where a variation is ordered then the contract is not invalidated but this may affect the contract price i.e. the amount of money actually paid. Note the use of the word 'forthwith'. Its meaning is not defined but it appears reasonable that if the Contractor confirms an oral order before the work is carried out and there is no rebuttal before it is carried out then it is deemed to be a bona fide ordered variation assuming that there is a reasonable period between delivery of the confirmation of variation ordered orally and the commencement of the execution of the work constituted therein.

A significant event in any contract is the actual final quantities compared with the billed quantities. Sub-Clause (3) dictates that there is no need to issue a variation as a result of changes in the actual quantities compared with those in the bill of quantities unless it is as a result of a variation issued under Sub-Clauses (1) and (2) above.

Unless there is an express term relating to work to be carried out during the period of maintenance, the Engineer cannot instruct its execution. Again the Contractor may choose to do so but on his terms.

If the Engineer cancels an element of the Works on the basis that the Employer's budget has been exceeded then that constitutes a breach and the Contractor would be entitled to a loss of profit, contribution to overheads and any other costs incurred by the Contractor as a result of this cancellation. However, where the Engineer substitutes work which achieves the same end result but is, in fact, cheaper then the Contractor would have no right of recovery.

Parties should bear in mind the ramifications emanating from the judgement arising from *Yorkshire Water Authority v. (Sir Alfred) McAlpine Ltd* (1985) 32 BLR 114. Hudson explains[7] that the construction of a tunnel was programmed in accordance with tender documents issued by the Owner (Employer), to be executed upstream. This was approved and accepted. Subsequently, the Contractor contended that this was impossible and, after delay had been incurred, proceeded to work downstream. The Contractor claimed a Variation under Clause 51(1). He held that upstream working was, in effect, a specified method of working, and under Clause 51(1) and by virtue of Clauses 13(1) and 13(3) he was entitled to a Variation, since the specified method of working was impossible.

The entire contents of this Clause are most important and all parties should apply its terms rigorously.

[Contractor's address]

[Engineer]

[Date]

Dear Sir

**[Contract description]**
**[Contract location]**
**[Specify area of work to which the oral order relates]**
**Confirmation of variation ordered orally by the engineer**

We confirm the variation ordered orally by you on [            ] to [            ]. We deem this to constitute a variation ordered orally pursuant to Clause 51(1) and (2) of the Conditions of Contract.

Yours faithfully


W Bell
Agent
for Unlimited Contracting Ltd


*Standard letter 23. Clause 51(1)/(2). Confirmation of variation ordered orally by the Engineer*

---

## 2.24. Clause 52

### SUB-CLAUSE (1)—VALUATION OF ORDERED VARIATIONS

The value of all variations ordered by the Engineer in accordance with Clause 51 shall be ascertained by the Engineer after consultation with the Contractor in accordance with the following principles. Where work is of similar character and executed under similar conditions to work priced in the Bill of Quantities it shall be valued at such rates and prices continued therein as may be applicable. Where work is not of a similar character or is not executed under similar conditions the rates and prices in the Bill of Quantities shall be used as the basis for valuation so far as may be reasonable failing which a fair valuation shall be made. Failing agreement between the Engineer and the Contractor as to any rate or price to be applied in the valuation of any variation the Engineer shall determine the rate or price in accordance with the foregoing principles and he shall notify the Contractor accordingly.

### SUB-CLAUSE (2)—ENGINEER TO FIX RATES

Provided that if the nature or amount of any variation relative to the nature or amount of the whole of the contract work or to any part thereof shall be such that in the opinion of the Engineer or the Contractor any rate or price contained in the Contract for any item of work is by reason of such variation rendered unreasonable or inapplicable either the Engineer shall give to the Contractor or the Contractor shall give to the Engineer notice before the varied work is commenced or as soon thereafter as is reasonable in all the circumstances that such rate or price should be varied and the Engineer shall fix such rate or price as in the circumstances he shall think reasonable and proper.

### SUB-CLAUSE (3)—DAYWORK

The Engineer may if in his opinion it is necessary or desirable order in writing that any additional or substituted work shall be executed on a daywork basis. The Contractor shall then be paid for such work under the conditions set out in the Daywork Schedule included in the Bill of Quantities and at the rates and prices affixed thereto by him in his Tender and failing the provision of a Daywork Schedule he shall be paid at the rates and prices and under the conditions contained in the 'Schedule of Dayworks carried out incidental to Contract Work' issued by the Federation of Civil Engineering Contractors current at the date of execution of the Daywork.

The Contractor shall furnish to the Engineer such receipts or other vouchers as may be necessary to prove the amounts paid and before ordering materials shall submit to the Engineer quotations for same for his approval.

In respect of all work executed on a daywork basis the Contractor shall during the continuance of such work deliver each day to the Engineer's Representative an exact list in duplicate of the names occupation and time of all workmen employed on such work and a statement also in duplicate showing the description and quantity of all materials and plant used thereon or therefor (other than plant which is included in the percentage addition in accordance with the Schedule under which payment for daywork is made). One copy of each list and statement will if correct or when agreed be signed by the Engineer's Representative and returned to the Contractor. At the end of each month the Contractor shall deliver to the Engineer's Representative a priced statement of the labour material and plant (except as aforesaid) used and the Contractor shall not be entitled to payment unless such lists and statements have been fully and punctually rendered. Provided always that if the Engineer shall consider that for any reason the sending of such list or statement by the Contractor in accordance with the foregoing provision was impracticable he shall nevertheless be entitled to authorised payment for such work either as daywork (on being satisfied as to the time employed and plant and materials used on such work) or at such value therefor as he shall consider fair and reasonable.

### SUB-CLAUSE (4)—NOTICE OF CLAIMS

(a) If the Contractor intends to claim a higher rate or price than one notified to him by the Engineer pursuant to sub-clauses (1) and (2) of this Clause or Clause 56(2) the Contractor shall within 28 days after such notification give notice in writing of his intention to the Engineer.

(b) If the Contractor intends to claim any additional payment pursuant to any Clause of these Conditions other than sub-clauses (1) and (2) of this Clause he shall give notice in writing of his intention to the Engineer as soon as reasonably possible after the happening of the events giving rise to the claim. Upon the happening of such events the Contractor shall keep such contemporary records as may reasonably be necessary to support any claim he may subsequently wish to make.

(c) Without necessarily admitting the Employer's liability the Engineer may upon receipt of a notice under this Clause instruct the Contractor to keep such contemporary records or further contemporary records as the case may be as are reasonable and may be material to the claim of which notice has been given and the Contractor shall keep such records. The Contractor shall permit the Engineer to inspect all records kept pursuant to this Clause and shall supply him with copies thereof as and when the Engineer shall so instruct.

(d) After the giving of a notice to the Engineer under this Clause the Contractor shall as soon as is reasonable in all the circumstances send to the Engineer a first interim account giving full and detailed particulars of the amount claimed to that date and of the grounds upon which the claim is based. Thereafter at such intervals as the Engineer may reasonably require the Contractor shall send to the Engineer further up to date accounts giving the accumulated total of the claim and any further grounds upon which it is based.

(e) If the Contractor fails to comply with any of the provisions of this Clause in respect of any claim which he shall seek to make then the Contractor shall be entitled to payment in respect thereof only to

# THE 5TH EDITION OF THE ICE CONDITIONS OF CONTRACT | 89

the extent that the Engineer has not been prevented from or substantially prejudiced by such failure in investigating the said claim.

(f) The Contractor shall be entitled to have included in any interim payment certified by the Engineer pursuant to Clause 60 such amount in respect of any claim as the Engineer may consider due to the Contractor provided that the Contractor shall have supplied sufficient particulars to enable the Engineer to determine the amount due. If such particulars are insufficient to substantiate the whole of the claim the Contractor shall be entitled to payment in respect of such part of the claim as the particulars may substantiate to the satisfaction of the Engineer.

This is a very important Clause.

Sub-Clause (1) specifies the means by which ordered variations, issued under Clause 51(1) and (2), are evaluated. The value is to be fixed by the Engineer after consultation with the Contractor in accordance with two principles. Where 'work is of similar character and executed under similar conditions to work priced in the Bill of Quantities it shall be valued at such rates and prices contained therein as may be applicable' or where 'work is not of a similar character or is not executed under similar conditions the rates and prices in the Bill of Quantities shall be used as the basis for valuation so far as may be reasonable failing which a fair valuation shall be made'. 'Failing agreement between the Engineer and Contractor' then the Engineer 'shall determine the rate or price in accordance with the foregoing principles and he shall notify the Contractor accordingly'. Note that the Engineer evaluates any variations 'after consultation with the Contractor'. The Contractor has the right to be consulted and neither the Engineer nor the Contractor should forget that fact. Notwithstanding, the Engineer may consult but pay scant regard to the Contractor's views and, indeed, may be right so to do. Note that for the work to be evaluated at billed rates both character and conditions must be similar to work priced in the bill of quantities—not one or the other.

It is often relatively simple for the Contractor to make a substantial case that ordered variations are not of similar character and/or are not executed under similar conditions. The duty placed on the Engineer is often rather a thankless task since he is unlikely to be privy to the method and details of how the Contractor's price was put together although some contracts require a breakdown of the Contractor's rates to be provided (see the amendment to Clause 14 contained in Part 4 of Section 1 of Volume 0: 'Model Contract

Document for Major Works and Implementation Requirements', the Model Contract Document for Wales).

Sub-Clause (2) allows the Engineer to set aside billed rates which may apply to any ordered variation if the nature or quantities warrant such action. Similarly the Contractor may apply for billed rates which may apply to any ordered variation to be set aside if the nature or quantities involved so warrant. The Engineer has the right to reduce or increase a billed rate if the nature or change in quantities justifies such action but it is rare for him to do so of his own volition. In contrast, it is common for the Contractor to suggest that billed rates are inappropriate and seek an increase.

The ordering of certain types of variation which are to be paid on a dayworks basis is empowered under Sub-Clause (3). The Engineer has the option of providing a dayworks schedule in the bill of quantities or employing the standard 'Schedule of Dayworks'[9] published by the Federation of Civil Engineering Contractors*. It is a common misconception amongst Engineers that the rates contained in the 'Schedule of Dayworks' published by the Federation of Civil Engineering Contractors is the civil engineering equivalent of owning a water company or running the national lottery. In cases where the Contractor is a small company then the Federation Dayworks Schedule's rates may offer a handsome return but in larger businesses which carry higher overhead levels, 'Schedule of Dayworks' rates may not cover costs. Nevertheless, under current pricing regimes they are often better than billed rates. When new rates are required on site (which happens often) it may be wise for the Contractor to evaluate the work on a dayworks basis and then translate the result into a rate for submission to the Engineer and justify this rate by stating the source of evaluation. This approach is frequently accepted and indeed the Engineer often has much difficulty in refuting such an approach since it is, after all, a true measure of the resources involved.

Some Employers have concluded that paying for items using the Federation Dayworks Schedule represents poor value for money. They are wrong. It is true that Works carried out under dayworks will usually cost more than they would have if they had been included in the original tender. At the time of writing, the labour cost multiplier is commonly around 2·5 in larger organisations

---

*The Federation of Civil Engineering Contractors ceased to exist on 15 November 1996. Several bodies exist which may fill the gap and it is assumed that one will produce a replacement for the Federation's 'Schedule of Dayworks' in due course.

(matching the 2·48 in the Federation schedule). In smaller businesses such an enhancement may well render a profit but in larger organisations, such as local authorities such a multiplier may well represent a loss.

In a vain bid to counter the costs incurred in executing elements by dayworks, some Employers devise contracts which purport to pay for Works which would normally attract an order for dayworks by some other means. Demonstrable costs is a current favourite. Typically, the Contractor is required to put a single percentage for labour, plant and materials (and this is the case in certain types of contracts let under the *Manual of Contract Documents for Highway Works*). This figure is then used as a multiplier which is applied to the actual costs paid. There are a number of points about this approach which warrant close consideration. The first point is that it is not possible to predict the mix of labour, plant and materials on any particular daywork. The second point is that plant costs are notional in that they do not relate to any fixed or even calculable cost. Often plant is charged internally at the whim of the accountant. Towards the end of the financial year, the overhead contribution attributable to plant may be reduced by a firm's management because usage has been higher than anticipated and hence the necessary contribution to overheads has been made. In pricing tenders, the Contractor's estimator may decide to reduce plant charges in order to secure the work. A nationally established list of plant rates is more fair. The reason why dayworks will often appear to return poor value for money is that the operation is poorly controlled. The fact is that the Contractor has little incentive to execute the work quickly. Indeed, it may be that the opposite is the case. The longer it takes, the more he gets paid. Furthermore, the Contractor may well be able to claim an extension of time on the basis of this (and other) dayworks items. Under dayworks it is important for the Engineer to exercise close control of the operation and the *ICE Conditions of Contract* afford that right.

The nature of civil engineering requires some means of evaluating payment on a cost plus basis. The 'Schedule of Dayworks' provides such a means.

Some Engineers will ask for new rates for items before ordering them or agreeing payment. In effect this is asking for a quotation. It is suggested that this may not always be in the interests of the Contractor who may wish to decline. After all, the tender has been awarded. There is no provision in the contract for the provision of quotations (other than for materials under the dayworks Clause 52(3)). Rates can be prepared on the basis of a daywork which is cashed up and then translated into a unitary rate. After all, surely that is how all rates are devised for the tender: by working out what the cost is going to be for the Contractor and adding an element of profit?

Sub-Clause (4) covers the procedures relating to notice of claims.

If the Contractor intends to claim a higher rate than that notified under Sub-Clauses (1) and (2) above or Clause 56 (2) ('Increase or Decrease of Rate') then the Engineer must be notified within 28 days of having advised the proposed rate. If the Contractor intends to claim additional payment under any other Clause of the *ICE Conditions of Contract* then the Engineer must be advised as soon as may be reasonable but not later than 28 days after the event and in such circumstances the Contractor must keep contemporary records to support the claim.

Upon receipt of a notice under this Clause then the Engineer may instruct the Contractor to keep such contemporary records as he thinks reasonable and material to supporting a claim. The Engineer has right of access to such records and the Contractor must supply copies upon request. The Contractor must supply a first interim account as soon as is reasonable in the circumstances and further accounts as the Engineer may require.

Where the Contractor does not comply with any part of this Clause then he is only entitled to the extent which has not been prejudiced by the Contractor's default. The Contractor is entitled to be paid the Engineer's assessment of any claims in accordance with Clause 60 ('Certificates and Payment').

Invariably the Engineer or the Engineer's staff will value variations at billed rates. In the vast majority of circumstances the work is not of similar character and/or is not executed under similar conditions. There are a large number of reasons why this is so; the work may be isolated in location, have to be executed out of sequence or there may be changes in elements of the work and so on. It is up to the Contractor to point out why billed rates are inapplicable. Contractors should resist the provision of rates before execution. The pricing risks are taken at the time of tender not as a result of some variation. It is better to value the work at dayworks rates (translated into a rate if the Engineer or the Engineer's Representative insist). Bear in mind that dayworks rates only apply where the work is incidental to the main work. Note also that after the issue of the Certificate of Completion the Contractor cannot be forced to accept dayworks rates. It may be that the Contractor is prepared to execute variations after the Certificate of Substantial Completion has been granted if a premium is paid but the method of ensuring adequate recompense rests with individual contractors.

During discussion of the various Clauses of the *ICE Conditions of Contract*, all the Clauses require notice under Clause 52(4)(a) or (b) ('Notice of Claims') and all the standard letters of claim in this book illustrate that point. It cannot be over-emphasised how important it is for a Contractor to establish his intention to claim even if he subsequently decides not to pursue any claim for whatever reason. Contractors should include in all valuations a sum (even if it is just an estimate) related to all claims. This has the effect of establishing with the Engineer that a claim has arisen and starts the (often over-long) settlement process.

---

[Contractor's address]

[Engineer]

[Date]

Dear Sir

**[Contract description]**
**[Contract location]**
**[Specify subject of ordered variation]**
**Request to vary rate**

We refer to the variation ordered under cover of your letter dated [         ] related to the above. We consider the [rate/price] contained in the contract (Item No. [         ]) to be inappropriate since the [nature/amount] of the variation relative to the [nature/amount] of the whole of the contract work is by reason of such variation rendered unreasonable or inapplicable. Accordingly we request that such [rate/price] be varied and ask you to fix a reasonable and proper [rate/price].

This notice is given pursuant to Clause 52(2) of the Conditions of Contract.

Yours faithfully


E James
Agent
for Unlimited Contracting Ltd

94 | CLAIMS ON HIGHWAY CONTRACTS

*Standard letter 24. Clause 52(2). Request for the Engineer to vary a rate/price for ordered variation*

[Contractor's address]

[Engineer]

[Date]

Dear Sir

**[Contract description]**
**[Contract location]**
**[Specify subject of notified rate or price]**
**Notice of intention to claim a higher rate than that notified for an ordered variation**

We refer to your valuation dated [          ] issued pursuant to Clause 52(1) of the Conditions of Contract and related to the variation ordered under cover of your letter dated [          ]. We intend to claim a higher [rate/price] than that notified pursuant to Clause 52(4)(a).

Yours faithfully


J Reed
Agent
for Unlimited Contracting Ltd


*Standard letter 25. Clause 52(4)(d). Notice of intention to claim for a higher rate/price than that notified*

[Contractor's address]

[Engineer]

[Date]

Dear Sir

**[Contract description]**
**[Contract location]**
**[Specify subject of claim]**
**Full and detailed particulars of claim**

We refer to our notice of claim dated [          ] issued pursuant to Clause 52(4)(a)/(b). Please find attached a first interim account giving full and detailed particulars of the amount claimed as at the above date.

Please advise us of the intervals at which you require further up to date accounts giving the accumulated total of the claim and any further grounds upon which it is based.

This submission is made pursuant to Clause 52(4)(d) of the Conditions of Contract.

Yours faithfully


S Terry
Agent
for Unlimited Contracting Ltd

Enc


*Standard letter 26. Clause 52(4)(d). First interim intimation of value of claim*

Standard letter 26. illustrates the type of letter which has to be sent to the Engineer having served a notice of claim. It should contain details of the claim and the grounds upon which it is based. Thereafter, these details should be updated and sent to the Engineer at intervals dictated by the Engineer and should also indicate any further grounds of claim related to the same claim and not to new claims.

## 2.25. Clause 55

### SUB-CLAUSE (1)—QUANTITIES

The quantities set out in the Bill of Quantities are the estimated quantities of the work but they are not to be taken as the actual and correct quantities of the Works to be executed by the Contractor in fulfilment of his obligations under the Contract.

### SUB-CLAUSE (2)—CORRECTION OF ERRORS

Any error in description in the Bill of Quantities or omission therefrom shall not vitiate the Contract nor release the Contractor from the execution of the whole or any part of the Works according to the Drawings and Specification or from any of his obligations or liabilities under the Contract. Any such error or omission shall be corrected by the Engineer and the value of the work actually carried out shall be ascertained in accordance with Clause 52. Provided that there shall be no rectification of any errors omissions or wrong estimates in the descriptions rates and prices inserted by the Contractor in the Bill of Quantities.

Sub-Clause (1) confirms that the quantities are estimates only.

Sub-Clause (2) means that the contract cannot be set aside or reduced because of errors of descriptions or omissions from the bill of quantities. These are to be corrected by the Engineer and valued as if they were variations in accordance with Clauses 51 and 52.

[Contractor's address]

[Engineer]

[Date]

Dear Sir

**[Contract description]**
**[Contract location]**
**[Specify subject of claim]**
**[Error/omission] from bill of quantities**
**Notice of intention to claim a higher rate than that notified and extension of time**

We refer to your letter dated [           ] which corrected [state details of the error/omission] pursuant to Clause 55(2) of the Conditions of Contract and contained your evaluation of the payment of [           ] issued pursuant to Clause 52(1) of the Conditions of Contract. We intend to claim a higher [rate/price] than that notified. Furthermore, we would advise you that this [error/omission] has caused a delay of [           ] [days/weeks]. Pursuant to Clause 44(1) of the Conditions of Contract, we attach a schedule showing full and detailed particulars thereof. Accordingly, we request that you grant an interim assessment of extension of time pursuant to Clause 44(2) of the Conditions of Contract.

This submission is made pursuant to Clause 52(4)(a) of the Conditions of Contract.

Yours faithfully


B McGhee
Agent
for Unlimited Contracting Ltd

Enc


*Standard letter 27. Clause 55(2). Notice of claim for increased rate and extension of time in relation to errors/omissions from bill of quantities.*

Standard letter 27. relates to the situation where the Engineer evaluates an error or omission under the terms of Clause 52 and the Contractor is unwilling to accept that valuation.

## 2.26. Clause 56

### SUB-CLAUSE (1)—MEASUREMENT AND VALUATION

The Engineer shall except as otherwise stated ascertain and determine by admeasurement the value in accordance with the Contract of the work done in accordance with the Contract.

### SUB-CLAUSE (2)—INCREASE OR DECREASE OF RATE

Should the actual quantities executed in respect of any item be greater or less than those stated in the Bill of Quantities and if in the opinion of the Engineer such increase or decrease of itself shall so warrant the Engineer shall after consultation with the Contractor determine an appropriate increase or decrease of any rates or prices rendered unreasonable or inapplicable in consequence thereof and shall notify the Contractor accordingly.

### SUB-CLAUSE (3)—ATTENDING FOR MEASUREMENT

The Engineer shall when he requires any part or parts of the work to be measured give reasonable notice to the Contractor who shall attend or send a qualified agent to assist the Engineer or the Engineer's Representative in making such measurement and shall furnish all particulars required by either of them. Should the Contractor not attend or neglect or omit to send such agent then the measurement made by the Engineer or approved by him shall be taken to be the correct measurement of the work.

Sub-Clause (1) establishes that the Engineer will fix the value of the contract by measuring the work done and paying the tendered amount.

Sub-Clause (2) permits the Engineer, after consultation with the Contractor, to increase or decrease billed rates in accordance with the actual quantities of work executed.

Sub-Clause (3) requires the Contractor to attend for measurement and if he fails to do so establishes the Engineer's measurement as that which will be used to evaluate the contract.

It is rare for the Engineer to reduce rates because quantities change. It is even more rare for the Engineer to increase rates because quantities change without prompting by the Contractor. The fact that an ordered variation is not required for changes in quantities should not inhibit the Contractor from making claims for alterations in rates where the circumstances so warrant. It is not possible to cite exactly when the level of change merits a claim. It may be relatively small. If, for example, the bill states 1000 m$^3$ of acceptable material and so the Contractor has only made provision for 1000 m$^3$ but the amount is actually 1050 m$^3$ then, if the Contractor has to haul it another 5 miles, a valid claim would result.

The Contractor has a right of contractual claim via Clause 52(4)(a).

---

[Contractor's address]

[Engineer]

[Date]

Dear Sir

**[Contract description]**
**[Contract location]**
**[Specify affected item/s]**
**Increase in rate due to change in quantity**

The actual amount of [            ] compared with the billed quantity of [            ] renders the tender rate inapplicable by reason of [            ]. We request that you exercise your authority under Clause 56(2) of the Conditions of Contract and agree to a re-rating of [Item Number/5].

We shall include a sum related hereto in the next monthly statement submitted in accordance with Clause 60(1).

Please note that this letter constitutes a notice required pursuant to Clause 52(4)(a) of the Conditions of Contract. Appropriate contemporary records will be kept to support an application for additional payment.

Yours faithfully

W Dixon
Agent
for Unlimited Contracting Ltd

*Standard letter 28. Clause 56(2). Request for re-rating as a result of a change in quantities*

---

[Contractor's address]

[Engineer]

[Date]

Dear Sir

**[Contract description]**
**[Contract location]**
**[Specify affected item/s]**
**Increase in rate due to change in quantity**

We refer to your letter dated [            ] concerning the actual amount of [            ] compared with the billed quantity of [            ] and the rate notified therein issued pursuant to Clause 56(2) of the Conditions of Contract. We consider this rate inappropriate because [            ] and we intend to claim a more appropriate rate in the next statement.

This letter constitutes a notice required pursuant to Clause 52(4)(a) of the Conditions of Contract. Appropriate contemporary records will be kept to support an application for additional payment and extension of time pursuant to Clause 51(3) and Clause 44(1) of the Conditions of Contract.

Yours faithfully

E James
Agent
for Unlimited Contracting Ltd

*Standard letter 29. Clause 56(2). Notice of claim as a result of a notification of rate related to a change in quantities*

Where there has been a significant change in quantities which warrants a request for an extension of time then, technically (see Clause 44—'other special circumstances of any kind whatsoever'), that request should be made in relation to Clause 51(3) of the *ICE Conditions of Contract* and this is illustrated in standard letter 29.

## 2.27. Clause 59A

### SUB-CLAUSE (1)—NOMINATED SUB-CONTRACTORS—OBJECTION TO NOMINATION

Subject to sub-clause (2)(c) of this Clause the Contractor shall not be under any obligation to enter into any sub-contract with any Nominated Sub-contractor against whom the Contractor may raise reasonable objection or who shall decline to enter into a sub-contract with the Contractor containing provisions

(a) that in respect of the work goods materials or services the subject of the sub-contract the Nominated Sub-contractor will undertake towards the Contractor such obligations and liabilities as will enable the Contractor to discharge his own obligations and liabilities towards the Employer under the terms of the Contract

(b) that the Nominated Sub-contractor will save harmless and indemnify the Contractor against all claims demands and proceedings damages costs charges and expenses whatsoever arising out of or in connection with any failure by the Nominated Sub-contractor to perform such obligations or fulfils such liabilities

(c) that the Nominated Sub-contractor will save harmless and indemnify the Contractor from and against any negligence by the Nominated Sub-contractor his agents workmen and servants and against any misuse by him or them of any Constructional Plant or Temporary Works provided by the Contractor for the purposes of the Contract and for all claims as aforesaid

(d) equivalent to those contained in Clause 63.

### SUB-CLAUSE (2)—ENGINEER'S ACTION UPON OBJECTION

If pursuant to sub-clause (1) of this Clause the contractor shall not be obliged to enter into a sub-contract with a Nominated Sub-contractor and shall decline to do so the Engineer shall do one or more of the following

(a) nominate an alternative sub-contractor in which case sub-clause (1) of this Clause shall apply;

(b) by order under Clause 51 vary the Works or the work goods materials or services the subject of the Provisional Sum or Prime Cost Item including if necessary the omission of any such work goods materials or services so that they may be provided by workmen contractors or suppliers as the case may be employed by the Employer either concurrently with the Works (in which case Clause 31 shall apply) or at some other date. Provided that in respect of the omission of any Prime Cost Item there shall be included in the Contract Price a sum in respect of the Contractor's charges and profit being a percentage of the estimated value of such work goods material or services omitted at the rate provided in the Bill of Quantities or inserted in the Appendix to the Form of Tender as the case may be;

(c) subject to the Employer's consent where the Contractor declines to enter into a contract with the Nominated Sub-contractor only on the grounds of unwillingness of the Nominated Sub-contractor to contract on the basis of the provisions contained in paragraphs (a) (b) (c) or (d) of sub-clause (1) of this Clause direct the Contractor to enter into a contract with the Nominated Sub-contractor on such other terms as the Engineer shall specify in which case sub-clause (3) of this Clause shall apply;

(d) in accordance with Clause 58 arrange for the Contractor to execute such work or to supply such goods materials or services.

### SUB-CLAUSE (3)—DIRECTION BY ENGINEER

If the Engineer shall direct the Contractor pursuant to sub-clause (2) of

# THE 5TH EDITION OF THE ICE CONDITIONS OF CONTRACT | 103

this Clause to enter into a sub-contract which does not contain all the provisions referred to in sub-clause (1) of this Clause

(a) the Contractor shall not be bound to discharge his obligations and liabilities under the Contract to the extent that the sub-contract terms so specified by the Engineer are inconsistent with the discharge of the same

(b) in the event of the Contractor incurring loss or expense or suffering damage arising out of the refusal of the Nominated Sub-contractor to accept such provisions the Contractor shall subject to Clause 52(4) be paid in accordance with Clause 60 the amount of such loss expense or damage as the Contractor could not reasonably avoid.

## SUB-CLAUSE (4)—CONTRACTOR RESPONSIBLE FOR NOMINATED SUB-CONTRACTS

Except as otherwise provided in this Clause and in Clause 59B the Contractor shall be as responsible for the work executed or goods materials or services supplied by a Nominated Sub-contractor employed by him as if he had himself executed such work or supplied such goods materials or services or had sub-let the same in accordance with Clause 4.

## SUB-CLAUSE (5)—PAYMENTS

For all work executed or goods materials or services supplied by Nominated Sub-contractors there shall be included in the Contract Price

(a) the actual price paid or due to be paid by the Contractor in accordance with the terms of the sub-contract (unless and to the extent that any such payment is the result of a default of the Contractor) net of all trade and other discounts rebates and allowances other than any discount obtainable by the Contractor for prompt payment

(b) the sum (if any) provided in the Bill of Quantities for labours in connection therewith or if ordered pursuant to Clause 58(7)(b) as may be determined by the Engineer

(c) in respect of all other charges and profit a sum being a percentage of the actual price paid or due to be paid calculated (where provision has been made in the Bill of Quantities for a rate to be set against the relevant item of prime cost) at the rate inserted by the Contractor against that item or (where no such provision has been made) at the rate inserted by the Contractor in the Appendix to the

Form of Tender as the percentage for adjustment of sums set against Prime Cost Items.

### SUB-CLAUSE (6)—BREACH OF SUB-CONTRACT

In the event that the Nominated Sub-contractor shall be in breach of the sub-contract which breach causes the Contractor to be in breach of contract the Employer shall not enforce any award of any arbitrator or judgement which he may obtain against the Contractor in respect of such breach of contract except to the extent that the Contractor may have been able to recover the amount thereof from the Sub-contractor. Provided always that if the Contractor shall not comply with Clause 59B(6) the Employer may enforce any such award or judgement in full.

Where the Employer wishes to use a particular sub-contractor, this sub-contractor would have to be nominated in the contract and a provisional sum or prime cost item set aside for payment. Despite this nomination, normally the Employer will have no contractual relationship with the nominated sub-contractor and therefore no right to damages against any sub-contractors. Right to damages for breach of contract would lie solely with the main Contractor. The reason for the inclusion of this Clause is to give the Employer some control (albeit indirectly) over nominated sub-contractors.

Sub-Clause (1) covers the situation where the Contractor has problems with the appointment of a nominated sub-contractor. The Contractor is not obliged to enter into a sub-contract under the following conditions.

(*a*) The Contractor raises reasonable objection to the sub-contractor.
(*b*) The sub-contractor refuses
    (i) to include in his contract with the Contractor provisions to be liable for his performance in relation to the contract for his activities
    (ii) to indemnify the Contractor in connection with the sub-contractor's default for operations or negligence by himself or his agents etc., or
    (iii) to include in his contract a forfeiture clause (see Clause 63—'Forfeiture').

Sub-Clause (2) empowers the Engineer to take one or more of the following actions

(*a*) nominate an alternative sub-contractor and, in such cases, Sub-Clause (1) applies

(b) order a variation pursuant to Clause 51
(c) if the Employer agrees, then alter the conditions which will apply to any such sub-contract thus permitting the Contractor to contract with the sub-contractor, or
(d) order the Contractor to execute the work where a provisional sum is specified in the contract or get the agreement of the Contractor to undertake the work where it is the subject of a prime cost item in the contract.

Note that where the Engineer varies the work such that a prime cost item is omitted then the Contractor is entitled to charges and profit in recognition of the fact that a source of income has been lost.

Sub-Clause (3) relates to the situation where the Engineer compels the Contractor to enter into a sub-contract where all of the specified conditions do not apply. In such circumstances then the Contractor

(a) does not have to discharge his contractual obligations to the extent that they are compromised by the alteration to the terms of the sub-contract
(b) is entitled to the recovery of any loss or expense incurred or damage suffered.

---

[Contractor's address]

[Engineer]

[Date]

Dear Sir

**[Contract description]**
**[Contract location]**
**[Variation to contract in relation to use of nominated sub-contractor]**
**Notice of loss of charges and profit**

We refer to the variation ordered under cover of your letter dated [            ] cancelling the prime cost item connected with the construction of [            ]. Pursuant to Clause 59A(2)(b) we seek the recovery of the charges and profit which would have been paid to us in relation to the execution of this work.

We shall include a sum in the next statement issued pursuant to Clause 60(1) and supply a detailed costing thereof. Please note that this letter constitutes a notice required pursuant to Clause 52(4)(b) of the Conditions of Contract.

Yours faithfully


L Richard
Agent
for Unlimited Contracting Ltd


*Standard letter 30. Clause 59A(2)(b). Notice of recovery of charges and profit related to the cancellation of a prime cost item*

[Contractor's address]

[Engineer]

[Date]

Dear Sir

**[Contract description]**
**[Contract location]**
**[Contract with named nominated sub-contractor]**
**Notice of loss or expense incurred/damage suffered**

We refer to the nomination of [          ] as a sub-contractor on this contract and in particular to their refusal to accept all the provisions specified in Clause 59A(1) of the Conditions of Contract and your instruction directing us to enter into a sub-contract containing terms which are different to those set out in Clause 59A(1) of the Conditions of Contract. This has resulted in us [incurring loss and expense/suffering damage] as a result of [          ] and pursuant to Clause 59A(3)(b) we seek recovery of all associated costs.

This letter constitutes a notice required pursuant to Clause 52(4)(a) of the Conditions of Contract. Appropriate contemporary records will be kept to support an application for additional payment.

Yours faithfully


D Ruffin
Agent
for Unlimited Contracting Ltd


*Standard letter 31. Clause 59A(3)(b). Notice of claim as a result of a nominated sub-contractor refusing to accept specified provisions*

---

## 2.28. Clause 59B

### SUB-CLAUSE (1)—FORFEITURE OF SUB-CONTRACT

Subject to Clause 59A(2)(c) the Contractor shall in every sub-contract with a Nominated Sub-contractor incorporate provisions equivalent to those provided in Clause 63 and such provisions are hereinafter referred to as 'the Forfeiture Clause'.

### SUB-CLAUSE (2)—TERMINATION OF SUB-CONTRACT

If any event arises which in the opinion of the Contractor would entitle the Contractor to exercise his right under the Forfeiture Clause (or in the event that there shall be no Forfeiture Clause in the sub-contract his right to treat the sub-contract as repudiated by the Nominated Sub-contractor) he shall at once notify the Engineer in writing and if he desires to exercise such right by such notice seek the Employer's consent to his so doing. The Engineer shall by notice in writing to the Contractor inform him whether or not the Employer does so consent and if the Engineer does not give notice withholding consent within 7 days of receipt of the Contractor's notice the Employer shall be deemed to have consented to the exercise of the said right. If notice is given by the Contractor to the Engineer under this sub-clause and has not been withdrawn then notwithstanding that the Contractor has not sought the Employer's consent as aforesaid the Engineer may with the Employer's consent direct the Contractor to give notice to the Nominated Sub-contractor expelling the Nominated Sub-contractor from the sub-contract Works pursuant to the Forfeiture Clause or rescinding the sub-contract as the case may be. Any such notice given to the Nominated Sub-contractor is hereinafter referred to as a notice enforcing forfeiture of the sub-contract.

### SUB-CLAUSE (3)—ENGINEER'S ACTION UPON TERMINATION

If the Contractor shall give a notice enforcing forfeiture of the sub-contract whether under and in accordance with the Forfeiture Clause in the sub-contract or in purported exercise of his right to treat the sub-contract as repudiated the Engineer shall do any one or more of the things described in paragraphs (a) (b) and (d) of Clause 59A(2).

### SUB-CLAUSE (4)—DELAY AND EXTRA COST

If a notice enforcing forfeiture of the sub-contract shall have been given with the consent of the Employer or by the direction of the Engineer or if it shall have been given without the Employer's consent in circumstances which entitled the Contractor to give such a notice

(a) there shall be included in the Contract Price

  (i) the value determined in accordance with Clause 52 of any work the Contractor may have executed or goods or materials he may have provided subsequent to the forfeiture taking effect and pursuant to the Engineer's direction

  (ii) such amount calculated in accordance with paragraph (a) of Clause 59A(5) as may be due in respect of any work goods materials or services provided by an alternative Nominated Sub-contractor together with reasonable sums for labours and for all other charges and profit as may be determined by the Engineer

  (iii) any such amount as may be due in respect of the forfeited sub-contract in accordance with Clause 59A(5)

(b) the Engineer shall take any delay to the completion of the Works consequent upon the forfeiture into account in determining any extension of time to which the Contractor is entitled under Clause 44 and the Contractor shall subject to Clause 52(4) be paid in accordance with Clause 60 the amount of any additional cost which he may have necessarily and properly incurred as a result of such delay

(c) the Employer shall subject to Clause 60(7) be entitled to recover from the Contractor upon the certificate of the Engineer issued in accordance with Clause 60(3)

(i) the amount by which the total sum to be included in the Contract Price pursuant to paragraphs (a) and (b) of this sub-clause exceeds the sum which would but for the forfeiture have been included in the Contract Price in respect of work materials goods and services done supplied or performed under the forfeited sub-contract

(ii) all such other loss expense and damage as the Employer may have suffered in consequence of the breach of the sub-contract

all of which are hereinafter collectively called 'the Employer's loss'.

Provided always that if the Contractor shall show that despite his having complied with sub-clause (6) of this Clause he has been unable to recover the whole or any part of the Employer's loss from the Sub-contractor or the Employer shall allow or (if he has already recovered the same from the Contractor) shall repay to the Contractor so much of the Employer's loss as was irrecoverable from the Sub-contractor except and to the extent that the same was irrecoverable by reason of some breach of the sub-contract or other default towards the Sub-contractor by the Contractor or except to the extent that any act or default of the Contractor may have caused or contributed to any of the Employer's loss. Any such repayment by the Employer shall carry interest at the rate stipulated in Clause 60(6) from the date of the recovery by the Employer from the Contractor of the sum repaid.

## SUB-CLAUSE (5)—TERMINATION WITHOUT CONSENT

If notice enforcing forfeiture of the sub-contract shall have been given without the consent of the Employer and in circumstances which did not entitle the Contractor to give such a notice

(a) there shall be included in the Contract Price in respect of the whole of the work covered by the Nominated Sub-contract only the amount that would have been payable to the Nominated Sub-contractor on due completion of the sub-contract had it not been terminated

(b) the Contractor shall not be entitled to any extension of time because of such termination nor to any additional expense incurred as a result of the work having been carried out and completed otherwise than by the said Sub-contractor

(c) the Employer shall be entitled to recover from the Contractor any additional expense he may incur beyond that which he would have incurred had the sub-contract not been terminated.

## SUB-CLAUSE (6)—RECOVERY OF EMPLOYER'S LOSS

The Contractor shall take all necessary steps and proceedings as may be required by the Employer to enforce the provisions of the sub-contract and/or all other rights and/or remedies available to him so as to recover the Employer's loss from the Sub-contractor. Except in the case where notice enforcing forfeiture of the sub-contract shall have been given without the consent of the Employer and in circumstances which did not entitle the Contractor to give such a notice the Employer shall pay to the Contractor so much of the reasonable costs and expenses of such steps and proceedings as are irrecoverable from the Sub-contractor provided that if the Contractor shall seek to recover by the same steps and proceedings any loss damage or expense additional to the Employer's loss the said irrecoverable costs and expenses shall be borne by the Contractor and the Employer in such proportions as may be fair in all the circumstances.

As part of the contract, according to Sub-Clause (1), between a Contractor and a nominated sub-contractor there has to be a Clause equivalent to Clause 63 ('Forfeiture') unless the requirement is relaxed by the Employer (Clause 59A(2)).

Under Sub-Clause (2), if any event causes the Contractor to decide to exercise the right of forfeiture under the forfeiture provision in the contract with a nominated sub-contractor (or to repudiate the contract if there is no forfeiture provision) then the Engineer must be notified in writing and the Employer's permission sought for such action. The Engineer must write advising whether the Employer has or has not given consent. If the Engineer does not do so within seven days of receipt of the Contractor's notice then that is deemed to be consent for the Contractor to exercise the right of forfeiture (or repudiation). In such cases the term applying to such a notice is a 'notice enforcing forfeiture'.

If the Contractor issues a notice enforcing forfeiture then the Engineer may take the action specified in sections (a), (b) or (d) of Clause 59A(2) as follows.

(*a*) Section (a) specifies that the Engineer may nominate an alternative sub-contractor and, in such cases, Sub-Clause (1) applies.
(*b*) Section (b) states that a variation may be ordered pursuant to Clause 51.
(*c*) Section (d) specifies that the Engineer may order the Contractor to execute the work where a provisional sum is specified in the contract or get the agreement of the Contractor to undertake the work where it is the subject of a prime cost item in the contract.

[Contractor's address]

[Engineer]

[Date]

Dear Sir

**[Contract description]**
**[Contract location]**
**[Contract with named nominated sub-contractor]**
**Notice of delay and additional cost as a result of the forfeiture of a contract with a nominated sub-contractor**

We refer to the forfeiture of the contract with [         ] as a nominated sub-contractor on this contract and we write to record that, as a result, we have been delayed by [         ] [days/weeks] and incurred additional cost.

We hereby request that pursuant to Clause 59B(4)(b) of the Conditions of Contract, you grant an extension of time under the terms of Clause 44(1) of the Conditions of Contract and certify payment of this extra cost under the terms of Clause 60 of the Conditions of Contract.

We shall include a sum related hereto in the next monthly statement submitted in accordance with Clause 60(1) and confirm that appropriate contemporary records have been kept to support this application for an extension of time and additional payment. Please note that this letter constitutes a notice required pursuant to Clause 52(4)(b) of the Conditions of Contract.

Yours faithfully


J Blues
Agent
for Unlimited Contracting Ltd


*Standard letter 32. Clause 59B(4)(b). Notice of claim for extension of time and recovery of additional cost incurred as a result of the forfeiture of a contract with a nominated sub-contractor*

## 2.29. Clause 60

### SUB-CLAUSE (1)—MONTHLY STATEMENTS

The Contractor shall submit to the Engineer at monthly intervals a statement (in such form if any as may be prescribed in the Specification) showing

(a) the estimated contract value of the Permanent Works executed up to the end of that month

(b) a list of any goods or materials delivered to the Site for but not yet incorporated in the Permanent Works and their value

(c) a list of any of those goods or materials listed in the Appendix to the Form of Tender which have not yet been delivered to the Site but of which the property has vested in the Employer pursuant to Clause 54 and their value

(d) the estimated amounts to which the Contractor considers himself entitled in connection with all other matters for which provision is made under the Contract including any Temporary Works or Constructional Plant for which separate amounts are included in the Bill of Quantities;

unless in the opinion of the Contractor such values and amounts together will not justify the issue of an interim certificate.

Amounts payable in respect of Nominated Sub-contractors are to be listed separately.

### SUB-CLAUSE (2)—MONTHLY PAYMENTS

Within 28 days of the date of delivery to the Engineer or Engineer's Representative in accordance with sub-clause (1) of this Clause of the Contractor's monthly statement the Engineer shall certify and the Employer shall pay to the Contractor (after deducting any previous payments on account)

(a) the amount which in the opinion of the Engineer on the basis of the monthly statement is due to the Contractor on account of sub-clause (1)(a) and (1)(d) of this Clause less a retention as provided in sub-clause (5) of this Clause

(b) such amounts (if any) as the Engineer may consider proper (but in no case exceeding the percentage of the value stated in the Appendix to the Form of Tender) in respect of (b) and (c) of sub-clause (1) of this Clause which amounts shall not be subject to a retention under sub-clause (4) of this Clause.

# THE 5TH EDITION OF THE ICE CONDITIONS OF CONTRACT | 113

The amounts certified in respect of Nominated Sub-contracts shall be shown separately in the certificate. The Engineer shall not be bound to issue an interim certificate for a sum less than that named in the Appendix to the Form of Tender.

## SUB-CLAUSE (3)—FINAL ACCOUNT

Not later than 3 months after the date of the Maintenance Certificate the Contractor shall submit to the Engineer a statement of final account and supporting documentation showing in detail the value in accordance with the Contract of the work done in accordance with the Contract together with all further sums which the Contractor considers to be due to him under the Contract up to the date of the Maintenance Certificate. Within 3 months after receipt of this final account and of all information reasonably required for its verification the Engineer shall issue a final certificate stating the amount which in his opinion is finally due under the Contract up to the date of the Maintenance Certificate and after giving credit to the Employer for all amounts previously paid by the Employer and for all sums to which the Employer is entitled under the Contract up to the date of the Maintenance Certificate the balance if any due from the Employer to the Contractor or from the Contractor to the Employer as the case may be. Such balance shall subject to Clause 47 be paid to or by the Contractor as the case may require within 28 days of the date of the certificate.

## SUB-CLAUSE (4)—RETENTION

The retention to be made pursuant to sub-clause (2)(a) of this Clause shall be a sum equal to 5 per cent of the amount due to the Contractor until a reserve shall have accumulated in the hand of the Employer up to the following limits

(a) where the Tender Total does not exceed £50 000 5 per cent of the Tender Total but not exceeding £1 500, or

(b) where the Tender Total exceeds £50 000 3 per cent of the Tender Total

except that the limit shall be reduced by the amount of any payment that shall have been made pursuant to sub-clause (5) of this Clause.

## SUB-CLAUSE (5)—PAYMENT OF RETENTION MONEY

(a) If the Engineer shall issue a Certificate of Completion in respect of any Section or part of the Works pursuant to Clause 48(2) or (3) there shall become due on the date of issue of such certificate and shall be paid to the Contractor within 14 days thereof a sum equal to $1\frac{1}{2}$ per cent of the amount due to the Contractor at that date in respect of each Section or part as certified for payment pursuant to sub-clause (2) of this Clause provided that any sum or sums paid under this sub-clause shall not exceed in aggregate one half of the retention money deducted in accordance with sub-clause (2)(a) of this Clause from the payments made to the Contractor at the date of issue by the Engineer of the aforesaid Certificate of Completion.

(b) One half of the retention money less any sums paid pursuant to sub-clause (5)(a) of this Clause shall be paid to the Contractor within 14 days after the date on which the Engineer shall have issued a Certificate of Completion for the whole of the Works pursuant to Clause 48(1).

(c) The other half of the retention money shall be paid to the Contractor within 14 days after the expiration of the Period of Maintenance notwithstanding that at such times there may be outstanding claims by the Contractor against the Employer. Provided that if at such time there remain to be executed by the Contractor any outstanding work referred to under Clause 48 or any works ordered during such period pursuant to Clauses 49 and 50 the Employer shall be entitled to withhold payment until the completion of such works of so much of the second half of the retention money as shall in the opinion of the Engineer represent the cost of the works so remaining to be executed.

Provided further that in the event of different maintenance periods having become applicable to different Sections or parts of the Works pursuant to Clause 48 the expression of 'expiration of the Period of Maintenance' shall for the purposes of this sub-clause be deemed to be the expiration of the latest of such periods.

### SUB-CLAUSE (6)—INTEREST ON OVERDUE PAYMENTS

In the event of failure by the Engineer to certify or the Employer to make payment in accordance with sub-clauses (2) (3) and (5) of this Clause the Employer shall pay to the Contractor interest upon any payment overdue thereunder at a rate equivalent to 2 per cent plus the minimum rate at which the Bank of England will lend to a discount house having access to the Discount Office of the Bank current on the date upon which such payment first becomes overdue. In the event of any variation in the said Minimum Lending Rate being announced whilst such payment remains overdue the interest payable to the Contractor for the period that such payment remains overdue shall be correspondingly varied from the date of each such variation.

## SUB-CLAUSE (7)—CORRECTION AND WITHHOLDING OF CERTIFICATES

The Engineer shall have power to omit from any certificate the value of any work done goods or materials supplied or services rendered with which he may for the time being be dissatisfied and for that purpose or for any other reason which to him may seem proper may by any certificate delete correct or modify any sum previously certified by him.

Provided always that

(a) the Engineer shall not in any interim certificate delete or reduce any sum previously certified in respect of work done goods or materials supplied or services rendered by a Nominated Sub-contractor if the Contractor shall have already paid or be bound to pay that sum to the Nominated Sub-contractor

(b) if the Engineer in the final certificate shall delete or reduce any sum previously certified in respect of work done goods or materials supplied or services rendered by a Nominated Sub-contractor which sum shall have been already paid by the Contractor to the Nominated Sub-contractor the Employer shall reimburse to the Contractor the amount of any sum overpaid by the Contractor to the Sub-contractor in accordance with the certificates issued under sub-clause (2) of this Clause which the Contractor despite compliance with Clause 59B(6) shall be unable to recover from the Nominated Sub-contractor together with interest thereon at the rate stated in Clause 60(6) from 28 days after the date of the final certificate issued under sub-clause (3) of this Clause until the date of such reimbursement.

## SUB-CLAUSE (8)—COPY CERTIFICATE FOR CONTRACTOR

Every certificate issued by the Engineer pursuant to this Clause shall be sent to the Employer and at the same time a copy thereof shall be sent to the Contractor.

Sub-Clause (1) establishes that at (calendar) monthly intervals the Contractor must submit to the Engineer a statement showing

(a) the estimated value of the work done including the amounts due to nominated sub-contractors (shown separately)
(b) the amount or estimated amount of any claims to which the Contractor feels entitled, and
(c) a list and the value of goods and materials either delivered to the

site or not yet delivered to the site but whose ownership has been vested in the Employer

unless the total amount due does not reach the minimum certificate value. The minimum certificate value is often not reached at the beginning or end of a contract. Despite the fact that this Sub-Clause does not recognise that fact, the Contractor should still submit a statement and try to persuade the Engineer that an interim certificate should be issued.

Under Sub-Clause (2) the Engineer has to certify and the Employer has to pay to the Contractor the amount of the statement considered due to the Contractor less the amount already paid less retention. Payment has to be made within 28 days of the 'date of delivery' of the statement to the Engineer or Engineer's Representative. The Engineer is to certify the amounts of (*a*), (*b*) and (*c*) above with (*a*) and (*b*) being subject to the deduction of retention and the amount payable for nominated sub-contracts shown separately in the certificate. Failure of the Employer to pay the sum due within 28 days attracts interest paid to the Contractor (Clause 60(6)).

Some Engineers will ask the Contractor to present a revised statement which reflects the Engineer's view of what should be paid. The Contractor is advised to strongly resist this request; unless an express condition has been included in the contract, the Engineer has no power so to do. Agreement by the Contractor to such a request will result in incurring extra work (i.e. preparing a new statement) for which there will be no reimbursement and is likely to lead to late payment. An equally important consideration is that it is the Engineer's duty, as seen throughout this examination of the various Clauses, to assess any extra costs incurred by the Contractor and deleting them from the statement merely puts off the process of settling these issues.

This Sub-Clause also allows the Engineer not to certify payment if the amount does not reach the minimum level specified in the Appendix to the Form of Tender unless the Certificate of Completion for the whole of the Works has been granted. It would be reasonable for the Engineer to ignore this minimum value at the beginning of a contract or towards the end where the amount of measurable work may be such that its value does not reach the specified minimum.

Sub-Clause (3) deals with the final account and Contractors (and Engineers) should be acutely aware of the timescales. The Maintenance Certificate is issued at the end of the period of maintenance (which is stated in the Appendix to the Form of Tender) providing

# THE 5TH EDITION OF THE ICE CONDITIONS OF CONTRACT | 117

all defects have been rectified. Not later than three months after the date of the Maintenance Certificate the Contractor must submit the final account. 'Within 3 months after receipt of this final account and of all information reasonably required for its verification the Engineer shall issue a final certificate stating the amount which in his opinion is finally due under the Contract' with the net amount to be paid within 28 days of the date of the certificate.

Figure 2.1. shows how the *5th Edition of the ICE Conditions of Contract* foresees how the payment issues should be completed. The period between the issue of the final account by the Contractor and the issue of the Final Certificate can be extended by the Engineer if it is considered that the information submitted as part of the final account does not constitute 'all information reasonably

***Issued by***

| Issue of Maintenance Certificate | Engineer |

↓ 3 Months

| Statement of final account | Contractor |

↓ 3 Months

| Issue of Final Certificate | Engineer |

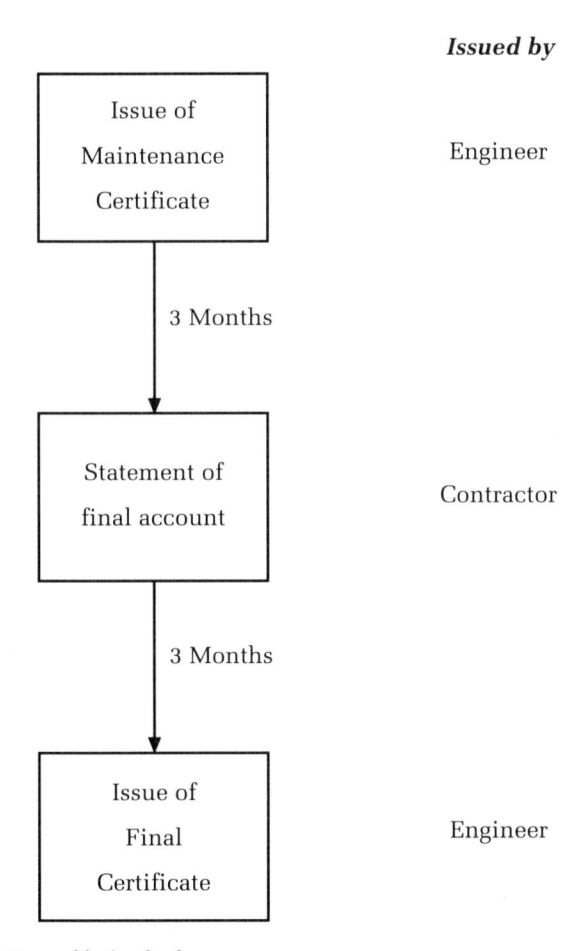

*Figure 2.1. Timetable for final account*

required for its verification'. The Contractor should respond to requests for further information timeously. Each time the Engineer asks for further information then the process is stopped. It starts again with the provision of the requested material by the Contractor. If, say two months later, the Engineer then requests further information then the clock stops, is reset to zero and only restarts when the request is met. This system should not be abused as it simply delays the inevitable adjudication and may force the Contractor to seek an Engineer's decision under Clause 66 which is not so restricted.

Sub-Clause (4) allows for retention with 5 per cent being deducted for all schemes limited to

(a) £1500 where the tender total is up to £50 000, or
(b) 3 per cent for Works costing above £50 000.

Sub-Clause (5) reduces the amount of retention after the Certificate of Completion for either a section or a part of the whole of the Works has been issued. In effect it reduces the retention to

(a) £750 where the tender total is up to £50 000, or
(b) $1\frac{1}{2}$ per cent for Works costing above £50 000.

This payment of the first element of the retention should be paid to the Contractor within 14 days of the Certificate of Completion being issued as dictated by the contract. Failure to do so attracts interest paid to the Contractor (Clause 60(6)).

The Employer may withhold that proportion of the outstanding retention which he decides (on the basis of the opinion of the Engineer) may be necessary to cover the cost of any remaining work. There is no requirement for the Contractor to request the release of retention money but he is advised to do so promptly; frequently the monies due are not paid timeously and equally many Contractors neglect to press for payment when it is due. Indeed, the fact that the Contractor does not ask for payment of retention money does not negate his right to interest for late payment of retention money.

Sub-Clause (6) dictates that failure to pay any monies due results in interest at a rate of 2 per cent above the Minimum Lending Rate (nowadays called the Base Lending Rate) of the Bank of England for the period of the default.

Sub-Clause (7) enables the Engineer to deduct the value of any work with which he is dissatisfied except where the work has been carried out by a nominated sub-contractor and sums have, in a previous certificate, been authorised.

Sub-Clause (8) requires the Engineer to send the certificate to the Employer with a copy to the Contractor.

It goes without saying that it is in the Contractor's interest to submit monthly statements on the same day every month without fail. Equally if the certified sum is not paid within 28 days then the Contractor should seek the payment of additional interest due. As explained previously, the (estimated) amount of claims should also be included. The amount due under the heading of materials on site can be considerable and should also be included. Note that three months after the issue of the Maintenance Certificate the final account must be submitted. Many Contractors present the final account well before that date. It is often wise not to label such accounts as final.

---

[Contractor's address]

[Engineer]

[Date]

Dear Sir

**[Contract description]**
**[Contract location]**
**Submission of monthly statement**

Please find attached our Statement Number [          ] which represents the work executed up to the end of [          ] for this contract.

This submission is made pursuant to Clause 60(1) of the Conditions of Contract.

Yours faithfully


E Blues
Agent
for Unlimited Contracting Ltd

Enc

*Standard letter 33. Clause 60(1). Submission of monthly statement*

[Contractor's address]

[Engineer]

[Date]

Dear Sir

**[Contract description]**
**[Contract location]**
**Submission of statement of final account**

Please find attached our statement of final account for this Contract including all information reasonably required for its verification.

This submission is made pursuant to Clause 60(3) of the Conditions of Contract.

Yours faithfully



D Dunn
Agent
for Unlimited Contracting Ltd

Enc


*Standard letter 34. Clause 60(3). Submission of final account*

---

2.30.   Clause 61

### SUB-CLAUSE (1)—MAINTENANCE CERTIFICATE

Upon the expiration of the Period of Maintenance or where there is more than one such period upon the expiration of the latest period and when all outstanding work referred to under Clause 48 and all work of repair amendment reconstruction rectification and making good of defects imperfections shrinkages and other faults referred to under Clauses 49 and 50 have been completed the Engineer shall issue to the

Employer (with a copy to the Contractor) a Maintenance Certificate stating the date on which the Contractor shall have completed his obligations to construct complete and maintain the Works to the Engineer's satisfaction.

### SUB-CLAUSE (2)—UNFULFILLED OBLIGATIONS

The issue of the Maintenance Certificate shall not be taken as relieving either the Contractor or the Employer from any liability one towards the other arising out of or in any way connected with the performance of their respective obligations under the Contract.

This Clause is self-explanatory but the Contractor should ensure that the Certificate contains the wording, the sense of which is equivalent to the Contractor having 'completed his obligations to construct complete and maintain the Works to the Engineer's satisfaction'. Use of the phrase 'pursuant to Clause 61(1)' would be acceptable.

It is very important that the Contractor ensures that this Certificate is issued at the earliest possible time by executing all items which are outstanding under Clauses 48, 49 and 50 and, if necessary, prompting the Engineer to do the same. The issue of the Maintenance Certificate starts the contractual clock on the provision of the final account.

---

[Contractor's address]

[Engineer]

[Date]

Dear Sir

**[Contract description]**
**[Contract location]**
**Issue of Maintenance Certificate**

We consider that with the expiry of the [latest] period of maintenance on [            ] and having completed all matters covered by Clauses 48, 49 and 50 you will wish to issue the Maintenance

Certificate stating the above date pursuant to Clause 61(1) of the Conditions of Contract.

Yours faithfully


H Sumlin
Agent
for Unlimited Contracting Ltd


*Standard letter 35. Clause 61(1). Request for Engineer to issue Maintenance Certificate*

---

## 2.31. Clause 66

### SUB-CLAUSE (1)—SETTLEMENT OF DISPUTES—ARBITRATION

If a dispute or difference of any kind whatsoever shall arise between the Employer and the Contractor in connection with or arising out of the Contract or the carrying out of the Works including any dispute as to any decision opinion instruction direction certificate or valuation of the Engineer (whether during the progress of the Works or after their completion and whether before or after the determination abandonment or breach of the Contract) it shall be referred in writing to and be settled by the Engineer who shall state his decision in writing and give notice of the same to the Employer and the Contractor.

### SUB-CLAUSE (2)—ENGINEER'S DECISION—EFFECT ON CONTRACTOR AND EMPLOYER

Unless the Contract shall have already been determined or abandoned the Contractor shall in every case continue to proceed with the Works with all due diligence and the Contractor and Employer shall both give effect forthwith to every such decision of the Engineer unless and until the same shall be revised by an arbitrator as hereinafter provided. Such decisions shall be final and binding upon the Contractor and the Employer unless and until the dispute or difference has been referred to arbitration as hereinafter provided and an award made and published.

## SUB-CLAUSE (3)—ARBITRATION—TIME FOR ENGINEER'S DECISION

(a) Where a Certificate of Completion of the whole of the Works has not been issued and

  (i) either the Employer or the Contractor be dissatisfied with any such decision of the Engineer, or

  (ii) the Engineer shall fail to give such decision for a period of one calendar month after such referral in writing

then either the Employer or the Contractor may within 3 calendar months after receiving notice of such decision or within 3 calendar months after the expiration of the said period of one month (as the case may be) refer the dispute or difference to the arbitration of a person to be agreed upon by the parties by giving notice to the other party.

(b) Where a Certificate of Completion of the whole of the Works has been issued and

  (i) either the Employer or the Contractor be dissatisfied with any such decision of the Engineer, or

  (ii) the Engineer shall fail to give such decision for a period of 3 calendar months after such referral in writing

then either the Employer or the Contractor may within 3 calendar months after receiving notice of such decision or within 3 calendar months after the expiration of the said period of 3 months (as the case may be) refer the dispute or difference to the arbitration of a person to be agreed upon by the parties by giving notice to the other party.

## SUB-CLAUSE (4)—PRESIDENT OR VICE-PRESIDENT TO ACT

(a) If the parties fail to appoint an arbitrator within one calendar month of either party serving on the other party a written Notice to Concur in the appointment of an arbitrator the dispute or difference shall be referred to a person to be appointed on the application of either party by the President for the time being of the Institution of Civil Engineers.

(b) If an arbitrator declines the appointment or after appointment is removed by order of a competent court or is incapable of acting or dies and the parties do not within one calendar month of the vacancy arising fill the vacancy then either party may apply to the President for the time being of the Institution of Civil Engineers to appoint another arbitrator to fill the vacancy.

(c) In any case where the President for the time being of the Institution of Civil Engineers is not able to exercise the functions conferred on him by this Clause the said functions may be exercised on his behalf by a Vice-President for the time being of the said Institution.

### SUB-CLAUSE (5)—ICE ARBITRATION PROCEDURE (1983)

(a) Any reference to arbitration shall be conducted in accordance with the Institution of Civil Engineers' Arbitration Procedure (1983) or any amendment or modification thereof being in force at the time of the appointment of the arbitrator. Such arbitrator shall have full power to open up review and revise any decision opinion instruction direction certificate or valuation of the Engineer and neither party shall be limited in the proceedings before such arbitrator to the evidence or arguments put before the Engineer for the purpose of obtaining his decision above referred to.

(b) Any such reference to arbitration shall be deemed to be a submission to arbitration within the meaning of the Arbitration Act 1950 or any statutory re-enactment or amendment thereof for the time being in force. The award of the arbitrator shall be binding on all parties.

(c) Any reference to arbitration may unless the parties otherwise agree in writing proceed notwithstanding that the Works are not then complete or alleged to be complete.

### SUB-CLAUSE (6)—ENGINEER AS WITNESS

No decision given by the Engineer in accordance with the foregoing provisions shall disqualify him from being called as a witness and giving evidence before the arbitrator on any matter whatsoever relevant to the dispute or difference so referred to the arbitrator as aforesaid.

Clause 66 deals with dispute resolution (but see the important note at the end of this Chapter on adjudication) and is the Clause which would be invoked in order to firstly obtain an Engineer's decision and secondly start the arbitration procedure.

Sub-Clause (1) states that if any dispute or difference whatsoever shall arise between the Employer and the Contractor including disputes as to any decision, opinion, instruction, direction, certificate or valuation of the Engineer then it shall be referred to and settled by the Engineer who shall state his decision in writing to both Employer and Contractor. So there must be an existing dispute or difference before the injured party can seek an Engineer's decision. The definition of dispute is very wide. The Engineer has to give a decision, or face the possibility of arbitration (see Sub-Clause (3)).

The Contractor should ensure that his letter states that he is seeking an Engineer's decision pursuant to Clause 66(1). The Engineer should respond stating that his decision is given pursuant to Clause 66(1). Where there is any doubt on either side then clarification should be sought in writing.

If the matter has not been referred to the Engineer for a decision then it falls outwith the procedure and technically cannot go to arbitration. However, this point may be academic because in order to make the matter open to arbitration all the Contractor has to do is refer it to the Engineer for his decision and after this procedure has been exhausted then it is open to arbitration. The net effect may be nil or it may then be the subject of a subsequent arbitration. The solution may be to include the matter within the terms of reference of the arbitrator.

Note that an Engineer's decision can be sought either during the contract or during the maintenance period or indeed thereafter limited only by the Engineer existing in terms of the contract. Normally the Engineer completes his appointment (becomes *functus officio* in legal terms) when he signs the Final Certificate unless the Contractor seeks an Engineer's decision under Clause 66(1).

Unless the contract has been abandoned or determined, Sub-Clause (2) requires the Contractor to continue with the Works and both the Contractor (and the Employer) to immediately implement the decisions of the Engineer unless and until an arbitrator makes an award reversing or revising this decision.

Sub-Clause (3) deals with the time available for the Engineer to give a decision.

Figure 2.2 illustrates the maximum time periods which are permissible under the *5th Edition* both for the Engineer to give his decision and for either the Contractor or Employer to refer the matter to the arbitration of an agreed party for his decision.

According to Sub-Clause (4), the arbitrator is agreed between the parties by one party serving on the other a (written) notice to concur but where agreement is not reached within one month of the notice then an arbitrator is appointed by the application by either party to the President of the Institution of Civil Engineers.

If the arbitrator

(*a*) declines the appointment
(*b*) after appointment is removed by order of a competent court

126 | CLAIMS ON HIGHWAY CONTRACTS

(c) is incapable of acting, or
(d) dies

and the parties do not agree the appointment of a replacement within one month, then by application by either party, a replacement shall be appointed by the President of the Institution of Civil Engineers to fill the vacancy.

Sub-Clause (5) deals with statutory provisions and states that once

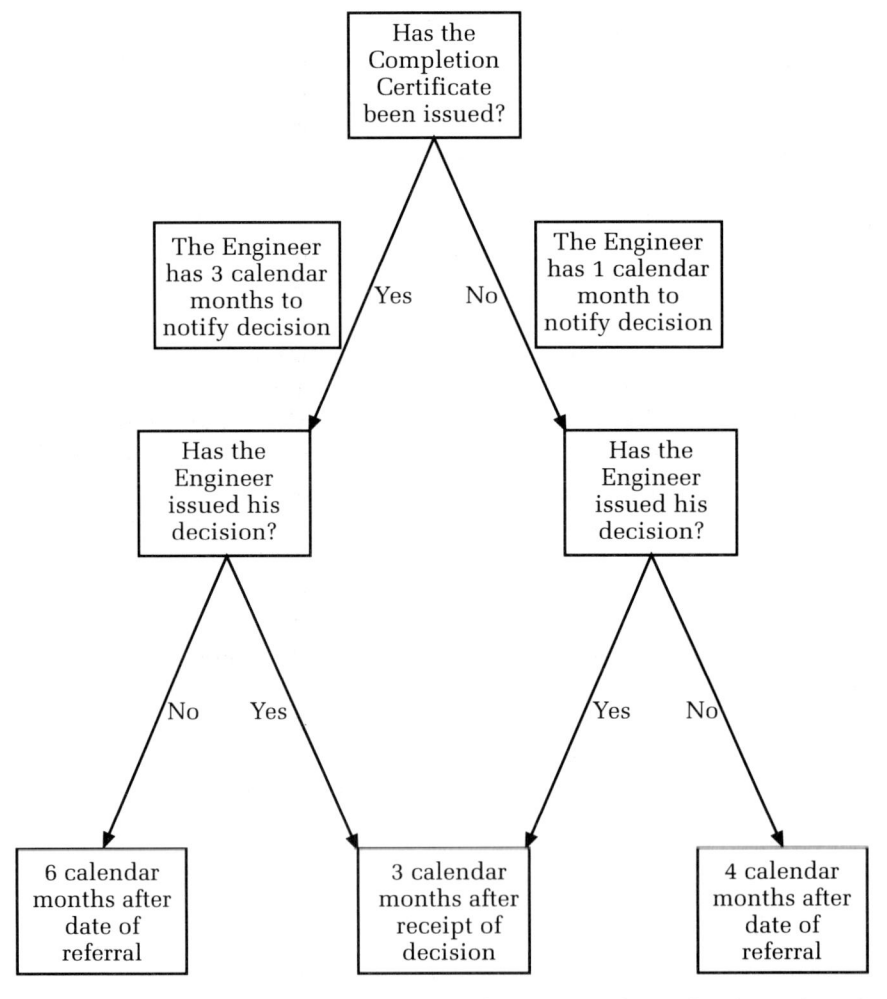

*Figure 2.2. Time periods before request for arbitration can be made after seeking the Engineer's decision*

a dispute is referred to arbitration in this form then the terms of one of the following apply

(*a*) the Arbitration Act 1950
(*b*) the Arbitration (Scotland) Act 1894, or
(*c*) any statutory re-enactment or amendment thereof being in force at the time of appointment of the arbitrator.

Any arbitration may be in accordance with the Institution of Civil Engineers' Arbitration Procedure (1983) as amended or modified and where the President appoints the arbitrator he may direct the use of this Procedure. The arbitrator may examine any matter relevant to the dispute and neither party is limited to the evidence placed before the Engineer at the time of his original decision which preceded the reference to arbitration. The arbitrator's award is stated as 'final and binding on the parties' but according to Abrahamson[5] the 'settlement of disputes cannot be given over completely to an arbitrator, who is always subject to some control by the courts, under the Arbitration Act'.

Sub-Clause (6) confirms that no decision of the Engineer disqualifies him from being called as a witness during the arbitration proceedings.

Both parties should think very carefully before going to arbitration. It is often a very expensive process. However, it is sometimes the only option, other than court action, which is available to an aggrieved party. In modern civil engineering, the Engineer is often an employee of the Employer and, rightly or wrongly but understandably, the Contractor feels that his decisions favour the Employer. It is sometimes the case that entrenched views are taken and that is why the matter ends up with an arbitrator. Both parties should seek advice from other independent parties before risking arbitration. Even those claims which Contractors feel are absolutely watertight can fail to find favour with an arbitrator. Equally, those claims which the Engineer feels are without substance can find themselves the subject of a favourable award. Contractors and Engineers should consult the legal profession once the possibility of arbitration becomes real.

## 2.32. Summary of provisions of Clauses

As each of the Clauses featured in this chapter have been considered, their provisions in relation to claims for extension of time or financial recompense have been identified. Table 2.1 identifies those

Table 2.1. (below, facing and overleaf). Clauses which feature a right to an extension of time and/or increased payment

| Clause no. and description | Entitlement | Eng. | Con. | Notes |
|---|---|---|---|---|
| 5—Documents mutually explanatory | Extension of time and reasonable cost | ✓ | | Clause 13(3) |
| 7(3)—Delay in issue of documents by Engineer | Extension of time and reasonable cost | ✓ | | Clauses 44(1) and 52(4)(b) |
| 12(3)—Adverse physical conditions and artificial obstructions not reasonably foreseeable by experienced contractor | (i) Extension of time<br>(ii) Reasonable cost plus profit for work and plant<br>(iii) Reasonable cost for delay or disruption | ✓<br>✓<br><br>✓ | | Clause 44(1)<br>If suspension order see Clause 40, if ordered variation see Clause 51<br>Clause 13(3) if Engineer issues instruction |
| 13(3)—Delay or disruption due to compliance with Engineer's instruction including those issued to clarify ambiguity or discrepancy not reasonably foreseeable by an experienced contractor | Extension of time and reasonable cost for delay or disruption | ✓ | | Clauses 44(1) and 52(4)(b) |
| 14(6)—Unreasonable delay in consent to proposed methods of construction or design criteria limitations not reasonably foreseeable by an experienced contractor | Extension of time and fair cost | ✓ | | Clauses 44(1) and 52(4)(b) |

# THE 5TH EDITION OF THE ICE CONDITIONS OF CONTRACT | 129

Table 2.1.—continued

| Clause no. and description | Entitlement | Eng. | Con. | Notes |
|---|---|---|---|---|
| 17—Error based on incorrect data supplied by Engineer or Engineer's Representative | (i) Cost of rectifying error<br>(ii) Extension of time and reasonable cost for delay or disruption | ✓ | ✓ | Clauses 52(4)(b)<br>Clause 13(3) if instruction given by Engineer |
| 20(2)—Damage loss or injury as a result of event which is defined as an excepted risk | (i) Expense of making good or repairing<br>(ii) Extension of time and reasonable cost for delay or disruption | ✓ | ✓ | Clause 52(4)(b)<br>Clause 13(3) if instruction given by Engineer |
| 27(6)—Variation ordering work which is emergency works, in controlled land or in a prospectively maintainable highway | Extension of time and additional cost | ✓ | | Clauses 44(1) and 52(4)(b) |
| 31(2)—Affording facilities for other contractors not reasonably foreseeable by an experienced contractor | Extension of time and reasonable cost | ✓ | | Clauses 44(1) and 52(4)(b) |
| 36(2)—Sampling not intended by or provided for in contract | (i) Cost of sampling and testing<br>(ii) Extension of time and reasonable cost for delay or disruption | ✓ | ✓ | Clause 52(4)(b)<br>Clause 13(3) if instruction given by Engineer |
| 38(2)—Making openings, reinstating and making good | (i) Cost borne by Employer<br>(ii) Extension of time and reasonable cost for delay or disruption | ✓ | ✓ | Clause 52(4)(b)<br>Clause 13(3) if instruction given by Engineer |

Table 2.1.—continued

| Clause no. and description | Entitlement | Eng. | Con. | Notes |
|---|---|---|---|---|
| 40(1)—Suspension not in contract or due to weather or not due to default of Contractor or necessary for proper execution of works of safety or due to default of Employer or Engineer or due to excepted risks | (i) Extension of time<br>(ii) Extra cost | ✓ | ✓ | Extension of time by Engineer—Clause 44(1)<br>Extra cost claim by Contractor—Clause 52(4)(b) |
| 42(1)—Failure of Employer to given possession of site | Extension of time and reasonable cost | ✓ |  | Clause 44(1) and 52(4)(b) |
| 44(1)—Delay due to variation ordered, increased quantities, any other cause mentioned in *Conditions*, exceptional adverse weather or other special circumstance | Extension of time | ✓ |  | Mechanism for claiming extension of time for all circumstances covered in these *Conditions* |
| 50—Carrying out searches, tests or trials on faults not liability of Contractor | Extension of time and reasonable cost for delay or disruption |  | ✓ | Clause 13 or 52(4)(b) |
| 52(2)—Evaluation of rate or price for variation | Reasonable or proper rate or price | ✓ | ✓ | Engineer notifies Contractor or Contractor notifies Engineer that billed rate inappropriate |
| 52(4)(a)—Notice of claim related to Engineer's assessment of rate or price for variation or change in billed quantity | Reasonable rate or price |  | ✓ | Only claim Clause for Contractor along with Clause 52(4)(b) |

## THE 5TH EDITION OF THE ICE CONDITIONS OF CONTRACT | 131

Table 2.1.—continued

| Clause no. and description | Entitlement | Eng. | Con. | Notes |
|---|---|---|---|---|
| 52(4)(b)—Notice of claim related to any other reason listed in the *Conditions* other than notified rate or price | As dictated by individual Clause | | ✓ | Primary claim Clause for Contractor |
| 55(2)—Error or omission in bill of quantities | (i) Value | ✓ | | Evaluated by Engineer under Clause 52—if Contractor unhappy then Clause 52(4)(a) claim |
| | (ii) Extension of time and reasonable cost of delay or disruption | ✓ | | Clause 13(3) if instruction given by Engineer |
| 56(2)—Alteration in rate due to actual quantities being different from those in bill of quantities | Reasonable rate or price | ✓ | | Evaluated by Engineer—if Contractor unhappy then Clause 52(4)(a) claim |
| 59A(3)(b)—Loss, expense or damage due to nominated sub-contractor refusing to accept specified conditions | Loss, expense or damage | | ✓ | Claim Clause 52(4)(b) |
| 59B(4)(a)(i)—Value of work, goods or materials executed by Contractor as a result of forfeiture of contract with nominated sub-contractor | Value of work, goods and material | ✓ | | Amount determined by Engineer under Clause 52—if Contractor unhappy then Clause 52(4)(a) claim |
| 59B(4)(b)—Delay due to forfeiture of contract with nominated sub-contractor | (i) Extension of time | ✓ | | Determined by Engineer per Clause 44(1) |
| | (ii) Additional cost | | ✓ | Claim by Contractor—Clause 52(4)(b) |

Clauses in the *5th Edition of the ICE Conditions of Contract* which have a provision for either or both. Those provisions which are ticked under the heading of 'Eng.' are those where the Engineer has the power to grant extension of time or payment or both as appropriate.

## 2.33. Summary of claims procedure

### 2.33.1. Introduction

The vast majority of situations where a Contractor is making a claim for extra payment will fall under only one of the Clauses in the *5th Edition of the ICE Conditions of Contract*. Some of the Clauses permit the Engineer to make allowance for various reasons for increased payment to the Contractor but it is always the Engineer who has the power to decide the validity of a request for extra payment. It is only when the Contractor disagrees with the Engineer's decision that a claim can be made. None of the Clauses entitles the Contractor to make a claim other than Clause 52(4) ('Notice of Claims'). Even Clause 12(1), ('Adverse Physical Conditions or Artificial Obstructions'), requires the Contractor to claim under Clause 52(4). Within any notice issued under this Clause, the Contractor should refer to the particular Clause which he feels has not been adequately addressed by the Engineer. The inclusion of a facility within various Clauses for the Engineer to make extra payment is a recognition that the Employer may be in breach of contract in some respect. These inclusions short cut what may be a valid claim under the law of contract. This explains why the tendency to re-frame certain contracts to avoid the use of standard forms such as the *ICE Conditions of Contract* is delusory. Unless they are prepared with great care then normal contract law considerations will prevail and the Contractor will still have a claim under those provisions.

### 2.33.2. What are costs?

The word 'cost' is defined in Clause 1(5) of the *Conditions of Contract* as being deemed to include overhead costs whether on or off site except where the contrary is expressly stated. Claims should therefore include an element of head office costs. On first view, only Clause 12 permits payment of cost plus 'a reasonable percentage addition thereto in respect of profit' for the extra work element.

However, where a Clause has a provision for work to be done at the expense of the Employer then it is suggested that this would include a reasonable element of profit; after all the Contractor would presumably have employed the resource elsewhere and would have, reasonably, earned profit. If the Contractor is not in default then why should he suffer? What constitutes cost can often be a function of how the Contractor plans or executes the work. Resources may well be deployed which will maximise income—for example, it may be financially more advantageous to employ hired plant or labour on items which are the subject of extra payment. The same is true of daywork operations where a quick calculation will determine the type of plant which renders the best return or indeed whether hired-in plant and labour give a better return. Against that must be weighed the effects on a delay or disruption claim. However, although there is a common law obligation to minimise costs in such circumstances, it is quite another matter to prove that this has or has not been done.

Note that claims made pursuant to Clause 12 may be more attractive than under some other Clauses because of the profit element therein although the Clause under which it will be made is largely a function of the nature of the event. How cost and profit are demonstrated within a claim is illustrated in Chapter 5.

A simplified version of Table 2.1 is given in Table 2.2 which lists those Clauses which contain a right to payment either directly or indirectly all, of course, related to the prevailing circumstances. It also suggests what the Contractor can reasonably include in any claim involving that Clause.

### 2.33.3. Claims for extensions of time

In claims terms, the granting of extensions of time is very important and can often be associated with considerable sums of money. Table 2.3 lists those Clauses under which the Engineer has the power to grant extensions of time. Where an extension of time is not granted to the extent that the Contractor feels adequately recognises the extent of delay then the Contractor must claim the extension under Clause 44(1).

## 2.34. Recent legislation to provide for adjudication

Sir Michael Latham's Report, *Constructing the team*,[10] was published in 1994 and resulted in the Housing Grants, Construction and

Table 2.2. (below, facing and overleaf). Clauses which feature a right to increased payment

| Clause no. and description | Entitlement | Quantum in claim |
|---|---|---|
| 5—Documents mutually explanatory | Reasonable cost | Cost including overhead contribution |
| 7(3)—Delay in issue of documents by Engineer | Reasonable cost | Cost including overhead contribution |
| 12(3)—Adverse physical conditions and artificial obstructions not reasonably forseeable by experienced contractor | (i) Reasonable cost plus profit for work and plant<br>(ii) Reasonable cost for delay or disruption | Cost including overhead contribution plus profit<br>Cost including overhead contribution |
| 13(3)—Delay or disruption due to compliance with Engineer's instruction including those issued to clarify ambiguity or discrepancy not reasonably foreseeable by an experienced contractor | Reasonable cost for delay or disruption | Cost including overhead contribution |
| 14(6)—Unreasonable delay in consent to proposed methods of construction or design criteria limitations not reasonably foreseeable by an experienced contractor | Fair cost | Cost including overhead contribution |
| 17—Error based on incorrect data supplied by Engineer or Engineer's Representative | Cost of rectifying error | Cost including overhead contribution plus profit |
| 20(2)—Damage loss or injury as a result of event which is defined as an excepted risk | (i) Expense of making good or repairing<br>(ii) Reasonable cost for delay or disruption | Cost including overhead contribution plus profit<br>Cost including overhead contribution |

Table 2.2.—continued

| Clause no. and description | Entitlement | Quantum in claim |
|---|---|---|
| 27(6)—Variation ordering work which is emergency works, in controlled land or in a prospectively maintainable highway | Additional cost | Cost including overhead contribution |
| 31(2)—Affording facilities for other contractors not reasonably foreseeable by an experienced contractor | Reasonable cost | Cost including overhead contribution |
| 36(2)—Sampling not intended by or provided for in contract | (i) Cost of sampling<br>(ii) Reasonable cost for delay or disruption | Cost including overhead contribution plus profit<br>Cost including overhead contribution |
| 36(3)—Testing not intended by or provided for in contract | (i) Cost of testing<br>(ii) Reasonable cost for delay or disruption | Cost including overhead contribution plus profit<br>Cost including overhead contribution |
| 38(2)—Making openings, reinstating and making good | (i) Cost borne by Employer<br>(ii) Reasonable cost for delay or disruption | Cost including overhead contribution plus profit<br>Cost including overhead contribution |
| 40(1)—Suspension not in contract or due to weather or not due to default of Contractor or necessary for proper execution of works or safety or due to default of Employer or Engineer or due to excepted risk | Extra cost | Cost including overhead contribution |

Table 2.2.—continued

| Clause no. and description | Entitlement | Quantum in claim |
|---|---|---|
| 42(1)—Failure of Employer to give possession of site | Reasonable cost | Cost including overhead contribution |
| 50—Carrying out searches, tests or trials on faults not liability of Contractor | Reasonable cost for delay or disruption | Cost including overhead contribution plus profit |
| 52(2)—Evaluation of rate or price for variation | Reasonable or proper rate or price | Cost including overhead contribution |
| 55(2)—Error or omission in bill of quantities | (i) Value<br>(ii) Reasonable cost of delay or disruption | Cost including overhead contribution<br>Cost including overhead contribution |
| 56(2)—Alteration in rate due to actual quantities being different from those in bill of quantities | Reasonable rate or price | Cost including overhead contribution plus profit |
| 59A(3)(b)—Loss, expense or damage due to nominated sub-contractor refusing to accept specified conditions | Loss, expense or damage | Cost including overhead contribution |
| 59B(4)(a)(i)—Value of work, goods or materials executed by Contractor as a result of forfeiture of contract with nominated sub-contractor | Value of work, goods and material | Cost including overhead contribution plus profit |
| 59B(4)(b)—Delay due to forfeiture of contract with nominated sub-contractor | Additional cost | Cost including overhead contribution |

# THE 5TH EDITION OF THE ICE CONDITIONS OF CONTRACT | 137

Table 2.3. (below and overleaf). Clauses which feature a right to an extension of time

| Clause No. and Description | Notes |
| --- | --- |
| 5—Documents mutually explanatory | Clause 13(3) |
| 7(3)—Delay in issue of documents by Engineer | Clause 44(1) |
| 12(3)—Adverse physical conditions and artificial obstructions not reasonably foreseeable by experienced contractor | Clause 44(1) |
| 13(3)—Delay or disruption due to compliance with Engineer's instruction including those issued to clarify ambiguity or discrepancy not reasonably foreseeable by an experienced contractor | Clause 44(1) |
| 14(6)—Unreasonable delay in consent to proposed methods of construction or design criteria limitations not reasonably foreseeable by an experienced contractor | Clause 44(1) |
| 17—Error based on incorrect data supplied by Engineer or Engineer's Representative | Clause 13(3) if instruction given by Engineer |
| 20(2)—Damage, loss or injury as a result of event which is defined as an as an excepted risk | Clause 13(3) if instruction given by Engineer |
| 27(6)—Variation ordering work which is emergency works, in controlled land or in a prospectively maintainable highway | Clause 44(1) |
| 31(2)—Affording facilities for other contractors not reasonably foreseeable by an experienced contractor | Clause 44(1) |
| 36(2)—Sampling not intended by or provided for in contract | Clause 13(3) if instruction given by Engineer |
| 36(3)—Testing not intended by or provided for in contract | Clause 13(3) if instruction given by Engineer |
| 38(2)—Making openings reinstating and making good | Clause 13(3) if instruction given by Engineer |
| 40(1)—Suspension not in contract or due to weather or not due to default of Contractor or necessary for proper execution of works or safety or due to default of Employer or Engineer or due to excepted risk | Clause 44(1) |

138 | CLAIMS ON HIGHWAY CONTRACTS

Table 2.3.—continued

| Clause No. and Description | Notes |
|---|---|
| 42(1)—Failure of Employer to give possession of site | Clause 44(1) |
| 44(1)—Delay due to variation ordered, increased quantities, any other cause mentioned in Conditions, exceptional adverse weather or other special circumstance | Mechanism for claiming for all circumstances covered in these Conditions |
| 50—Carrying out searches tests or trials on faults not liability of Contractor | Clause 13(3) if instruction given by Engineer |
| 55(2)—Error or omission in bill of quantities | Clause 13(3) if instruction given by Engineer |
| 59B(4)(b)—Delay due to forfeiture of contract with nominated sub-contractor | Clause 44(1) |

Regeneration Act which was given royal assent on 24 July 1996. Part II deals with construction contracts and may significantly affect the way that disputes are resolved in Civil Engineering. Basically it forces the parties to go to adjudication when requested by one of the parties to a construction contract. It will not come into force until a 'scheme' is issued by the Government. This is currently expected to be early 1998.

Details of this piece of legislation can be found in Raymond Joyce's excellent text[11] but basically it requires construction contracts to provide a timetable with the object of securing the appointment of the adjudicator and referral of the dispute to him within 7 days of one party referring a dispute for adjudication. Furthermore it requires the adjudicator to reach a decision within 28 days (although this period can be extended by agreement between the parties). It is hoped that this legislation will result in disputes reaching a conclusion at a much earlier stage than has been the case in the past.

## 2.35. Amendments to the *ICE Conditions of Contract*

At the time this book went to press, it became known that amendments to the *6th Edition of the ICE Conditions of Contract* were being considered.

Possible amendments include the following clauses: Clause 1 (Definitions); Clause 2 (Engineer and Engineer's Representative; Clause 60 (Certificates and Payments), Clause 66 (Settlement of Disputes); and Clause 69 (Tax Matters). There may be a new Clause 72 (Housing Grants, Construction and Regeneration Act 1996).

If these go ahead, then it is likely that similar changes will be made to the *5th Edition of the ICE Conditions of Contract*, and users must always ensure that they operate the correct contractual provisions.

## 2.36. References

1. **Duncan Wallace I. N.** *Hudson's Building and Civil Engineering Contracts*. Sweet & Maxwell, London, 1995, 11th edn, **1.009**, **1.224**, **4.042**, **7.013**.
2. **Duncan Wallace I. N.** Ibid., **1.009**.
3. **Duncan Wallace I. N.** Ibid., **8.063**.
4. **Furmston M.** *Contractors' Guide to the ICE Conditions of Contract, 5th Edition*. IPC Building & Contract Journals, Sutton, 1980, **5**.
5. **Abrahamson M. W.** *Engineering Law and the ICE Contracts*. E & F N Spon, London, 1979, 4th edn, **66**.
6. **Duncan Wallace I. N.** *Hudson's Building and Engineering Contracts*. Sweet & Maxwell, London, 1995, **7.032**.
7. **Abrahamson M. W.** *Engineering Law and the ICE Contracts*. E & F N Spon, London, 1979, 4th edn, **371–372**.
8. **Duncan Wallace I. N.** *Hudson's Building and Engineering Contracts*. Sweet & Maxwell, London, 1995, 11th edn, **10.024**.
9. **The Federation of Civil Engineering Contractors**. *Schedule Of Dayworks Carried Out Incidental To Contract Work*. The Federation Of Civil Engineering Contractors, London, 1990.
10. **Sir Michael Latham.** *Constructing the team*. HMSO, London, 1994.
11. **Joyce R.** *A commentary on construction contracts. Part II of the Housing Grants, Construction and Regeneration Act 1996*. Thomas Telford, London, 1996.

# 3

# The *7th Edition of the Manual of Contract Documents for Highway Works*: its constituents and functions

## 3.1. Preamble

The *Manual of Contract Documents for Highway Works* is most commonly, but incorrectly, referred to as the 'Specification for Highway Works'. This is probably due to the fact that when originally published there was only one document, the specification. The *Manual of Contract Documents for Highway Works*, currently consists of many discrete parts and is often referred to as 'the MCD'.

The current edition of the Manual is designated the 7th Edition but some elements have not appeared in seven editions and, indeed, some have never been published before.

Money and Hodgson's excellent books on the *Manual of Contract Documents for Highway Works*[1,2] contain a detailed examination of Volume 1: 'Specification for Highway Works' and also some notes on the 'Method of Measurement for Highway Works'.

The current edition of the *Manual of Contract Documents for Highway Works* (incorporating the August 1993 and August 1994 amendments) forms part of the subject of this book and this Chapter looks at the principles which pervade the production of the constituent elements particularly those which are relevant to claims. Where a particular section or part is unlikely to have any bearing in a claims situation then it is not discussed. It also includes a section on non-standard or rogue items and the pitfalls in their production which may result in claims.

Understanding these documents and the rules contained therein is vital if Contractors are to effectively implement the claims provision in the *5th Edition of the ICE Conditions of Contract*. Equally, if Engineers wish to minimise the possibility of claims then the rules which govern changes to the specification and the production of the bill of quantities and amendments to the method of measurement must be clearly understood. This Chapter sets out to meet that aim.

## 3.2. Introduction

The *7th Edition of the Manual of Contract Documents for Highway Works* consists of seven volumes, designated Volumes 0 to 6. Volumes 0, 3, 4, 5 and 6 contain several sections. The complete list of documents which comprise the *Manual of Contract Documents for Highway Works* is as follows.

**Volume 0: Manual of Contract Documents for Major Works and Implementation Requirements**
   Section 0: Introduction of Manual System
      Part 1 SA4/95 The Introduction of the Manual of Contract Documents for Highway Works
      Part 2 Volume Contents Pages and Volume Index
   Section 1: Model Contract Document for Highway Works
      Part 1 SD5/92 Implementation of the Model Contract Document for Highway Works
      Part 2 Model Contract Document for Highway Works—England February 1994 Edition
      Part 3 Model Contract Document for Highway Works—Scotland
      Part 4 Model Contract Document for Highway Works—Wales February 1994 Edition
      Part 5 Model Contract Document for Highway Works—Northern Ireland
   Section 2: Implementing Standards
      Part 1 SD1/92 Implementation of Specification for Highway Works and Notes for Guidance
      Part 2 SD2/92 Implementation of Highway Construction Details
      Part 3 SD3/92 Preparation of Bills of Quantities for Highway Works
      Part 4 Procedures for Adoption of Proprietary Manufactured Structures
      Part 5 SD6/95 Implementation of 1994 Annual Amendment to Specification for Highway Works and Notes for Guidance, Highway Construction Details and Preparation of Bills of Quantities for Highway Works
   Section 3: Advice Notes
      Part 1 SA1/95 Lists of Approved/Registered Products
      Part 2 SA2/92 Assessing Equivalence

Part 3 SA3/93 Testing in Highway Construction Contracts
Part 4 SA5/92 Radio Scheme for Road Construction and Maintenance Projects
Part 5 SA6/92 Introduction of 'Planning for Safety'
**Volume 1: Specification for Highway Works**
**Volume 2: Notes for Guidance on the Specification for Highway Works**
**Volume 3: Highway Construction Details**
Section 1: Carriageway and Other Details
Section 2: Safety Fences and Barriers
**Volume 4: Bills of Quantities for Highway Works**
Section 1: Method of Measurement for Highway Works
Section 2: Notes for Guidance on the Method of Measurement for Highway Works
Section 3: Library of Standard Item Descriptions for Highway Works
**Volume 5: Contract Documents for Specialist Activities**
Section 1: Geodetic Surveys
Part 1 Implementing Standard for the Specification of Geodetic Surveying Services
Section 2: Maintenance Painting of Steel Highway Structures
Part 1 Model Contract Document
Part 2 Standard SD7/94
Part 3 Advice Note SA7/94
Part 4 Specification
Part 5 Notes for Guidance on the Specification
Part 6 Bills of Quantities
**Volume 6: Departmental Standards and Advice Notes on Contract Documentation and Site Supervision**
Section 1: Standards
Part 1 SD10/95 Construction (Design and Management) Regulations 1994: Requirements for Health and Safety Plan
Part 2 SD11/95 Construction (Design and Management) Regulations 1994: Requirements for Health and Safety File
Section 2: Advice Notes
Part 1 SA8/94 Use of Substances Hazardous to Health in Highway Construction
Part 2 SA8/94 Use of Substances Hazardous to Health in Highway Construction Amendment October 1994

All these documents are published by HMSO and those shown above are those published up to mid-1996. HMSO does not publish the version of Part 3 of Section 1 of Volume 0: 'Model Contract Document for Major Works and Implementation Requirements', currently used as the model for Scotland; at the time of writing (June 1997) it is currently only available via The Scottish Office. Where a part is shown above it is available for separate purchase. Where a section is shown and is not divided into parts then the entire

## 7TH EDITION OF THE MANUAL OF CONTRACT DOCUMENTS | 143

section is available as a single purchase e.g. Volume 5, Section 1 Geodetic Surveys is available as a unit but consists of four parts as follows.

> Part 1 SD12/96: Implementing Standard for the Specification for Geodetic Surveying Services
> Part 2: Specification for Geodetic Surveying Services
> Part 3: Notes for Guidance on the Specification for Geodetic Surveying Services
> Part 4: Bill of Quantities for the Specification for Geodetic Surveying Services

Finally, at the time of writing Volumes 1, 2, 3 and 4 have been amended twice, in August 1993 and August 1994. Consequently, contracts may employ documents dated December 1991, August 1993 or August 1994.* This is clarified by reference to the schedules of pages and relevant publication dates which are included within the contract documentation as explained later. Volumes 1 and 2 are currently available separately as combined December 1991/August 1993 consolidated editions. Where the August 1994 versions apply only the amendments are available which requires the removal of some pages and replacement by those contained within the published amendment. In the cases of both Volumes 3 and 4, no consolidated versions have yet been made available. It does take a little work to understand the system and the inter-relationship of all the constituent elements. However, there is no doubt that it is a superb system. The one cautionary note which some may raise is to question whether such a substantial increase in the details of the specification over previous editions results in an equivalent improvement in the quality of the finished product without a disproportionate increase in cost.

### 3.3. Volume 0: 'Model Contract Document for Major Works and Implementation Requirements'

Volume 0 establishes how the system designated the *Manual of Contract Documents for Highway Works* is intended to operate. Volume 0 consists of four sections (Sections 0 to 3 inclusive).

---

*In fact, the Manual, though dated, was not published until July 1992, and the amendments, dated August 1993 and August 1994, were not published until July 1994 and December 1995 respectively.

### 3.3.1. Section 0: Introduction of Manual System

*Sub-Clause 1.1*  This establishes that the 'overseeing organisations' of England, Scotland, Wales and Northern Ireland (The Highways Agency, The Scottish Office Industry Department Roads Directorate, The Welsh Office Highways Directorate and The Department of the Environment for Northern Ireland, Roads Service respectively) have a comprehensive loose-leaf system of manuals (i.e. the *Manual of Contract Documents for Highway Works*) which form the primary documents required for the preparation of contracts for the construction, improvement or maintenance of trunk roads including motorways.

*Sub-Clause 1.8*  This mentions the *Manual of Contract Documents for Highway Works*' companion system for road and bridge design, the *Design Manual for Roads and Bridges*, which is published in a back-breaking (and wallet-breaking) 14 volumes (15 volumes in Scotland) which are contained in some 21 binders (22 binders in Scotland).

*Sub-Clauses 1.9 and 1.10*  These specify how the contents pages for Volumes 0, 5 and 6 and the index and contents pages for Volumes 1, 2, 3 and 4 respectively are handled.

*Sub-Clause 1.11*  This deals with the issue of amendments to Volumes 0, 5 and 6 and explains that the *Manual of Contract Documents for Highway Works* will be updated to match the latest edition of the 'Specification for Highway Works' with amendments being issued as looseleaf sheets for insertion at the appropriate points.

*Sub-Clause 1.13*  This refers to the Model Contract Documents, there being one for each of the overseeing departments preceded by an explanatory document on their usage denoted SD5. The full set is as follows.

- Volume 0 Section 1: Part 1 SD5/92
- Volume 0 Section 1: Part 2 Model Contract Document—England
- Volume 0 Section 1: Part 3 Model Contract Document—Scotland
- Volume 0 Section 1: Part 4 Model Contract Document—Wales
- Volume 0 Section 1: Part 5 Model Contract Document—Northern Ireland

*Sub-Clause 1.14* This deals with the issue of amendments to Volumes 1, 2, 3 and 4 and explains how these amendments will be issued and their format.

*Sub-Clause 1.15* This explains how control is achieved through supplying replacement text pages and a revised index. A revised 'Schedule of Pages and Relevant Publication Dates' will also be provided and the Sub-Clause instructs that these are to be 'reproduced unaltered and bound in the Specification and Bill of Quantities'.

*Sub-Clause 1.19* Suggestions as to how sections within the *Manual of Contract Documents for Highway Works* e.g. MCHW 0.2.3 meaning the *Manual of Contract Documents for Highway Works*, Volume 0, Section 2, Part 3 and MCHW 1.104 meaning *Manual of Contract Documents for Highway Works*, Volume 1, Clause 104 are made. (In this book the full description is used to avoid any ambiguity. The need for a shorthand reference is very desirable but unclear—is MCD the *Manual of Contract Documents* or the Model Contract Document?

*3.3.2. Volume 0, Section 1: Model Contract Document for Highway Works*

*Part 1. SD5/92 Implementation of the Model Contract Document for Highway Works* This explains how Parts 1, 2, 3 and 4 (Model Contract Documents for England, Scotland, Wales and Northern Ireland) are compiled. Each of these parts is divided into two sections with the first being the 'Instructions for Tendering' which are to be sent out with the tenders and the second being actual or typical contract documentation and is to be incorporated in full as appropriate in the specific contract document. Note that the second section contains mandatory changes to the *5th Edition of the ICE Conditions of Contract* (June 1973; revised January 1979; reprinted January 1986). This part also explains that the text is printed on the right hand page with guidance notes on the left hand page opposite.

*Part 2. Model Contract Document for Highway Works—England* Part 2 is the Model Contract for England. The Model Contract is divided into two sections. The first is the 'Instructions for Tendering' along with a 'Location and Description of the Works' and must not be bound into the tender documents since it does not form part of the contract documentation. There is however an acknowledgement that certain information may have a contractual significance. The second is the contractual part of the Model and contains actual or illustrative text which will form part of the contract document itself. The format of this document is such that the main text is printed on the right hand pages and what are called the 'guidance notes' are printed on the left hand pages. There are not always guidance notes for the Model Contract.

*Instructions for Tendering.* Clause 8 deals with the time for completion for the whole of the Works and, if applicable, the time/s for any specified sections (see Clause 43 of the *5th Edition of the ICE Conditions of Contract*). Contractors should always consider whether the time permitted is reasonable. There is no right of appeal, no matter how unreasonable the period may be. It may be that the Contractor plans to over-run and, accordingly, allows in his rates for the cost of liquidated damages.

Clause 16 deals with liquidated damages. The basis upon which they are calculated is stated. Their value is based on the price of the submitted tender and is calculated as 15 per cent of the tender price and then divided by 365 to give the daily liquidated damages figure. It is inserted into the Appendix to the Form of Tender immediately prior to the award of the tender. The notes for guidance (on the left hand side of the page) explain that this is the maximum which can normally apply. It states that 2·5 per cent of the 15 per cent is the allowance for site supervision and where this is exceeded then the liquidated damages should be calculated as 12·5 per cent of the final tender price with a daily supervision cost of £[       ] per day (note that the Highways Agency has chosen per day as the unit in order to maximise damages). This alternative formula must be stated in the Instructions for Tendering. The same method seems to apply regardless of economic conditions but it does adopt the formulaic approach which finds favour with the courts and, consequently, an action to set it aside may be unsuccessful albeit that it is not a particularly accurate means of loss assessment. There is also the possibility of damages being applied in the normal way and the sum so fixed exceeding that due under liquidated damages. It is far easier and probably more profitable to ensure that extensions of time are awarded.

Where the contract allows for sectional completion then the figures attributable to various sections are to be stated as percentages of the total liquidated damages daily value in the Appendix to the Form of Tender. As was seen in Chapter 2, there is nothing sacrosanct about the sections listed in the contract and if the Contractor can make a case for the award of a Certificate of Completion for a part of the Works then he should do so.

Clause 20 gives the period in weeks ('some ... weeks') which, it is anticipated, will elapse between the date of award of the contract and the Date for Commencement of the Works (see Clause 41 of the *5th Edition of the ICE Conditions of Contract*). This may produce a claim if it is substantially exceeded; on the other hand it may be that in contracts which contain a Price Fluctuation Clause (see the Contract Price Fluctuations Clause, called a special condition, in the *5th Edition of the Conditions of Contract*).*

*Model Contract Documentation.* This section starts with a suggested cover layout and contents page. These are followed by a number of documents which form part of the tender, 'Anti-Collusion Certificate', 'Prompt Payment Certificate', 'Forms of Agreement by Deed' and 'Form of Bond'. Thereafter is the section which amends the *5th Edition of the ICE Conditions of Contract* and the following notes relate to changes to any of the Clauses featured in Chapter 2 or any other Clause which may be relevant to claims.

The scope of Clause 7 is extended by the addition of the word 'specifications' after the word 'drawings' in Sub-Clauses 7(1) and (3). The standard Clause talks about drawings or instructions, the amended Clause now talks about drawings, specifications or instructions.

Clause 27(1) is altered to reflect the replacement of the Public Utilities Street Works Act 1950 by the New Roads and Street Works Act 1991 by substituting a new Sub-Clause (4) which reflects the changes to certain defined terms. It does not alter the basic sense and rights of the Contractor (nor of the Employer or the Engineer) contained within the original Clause.

Clause 42 is amended. The version featured in the *5th Edition of the ICE Conditions of Contract* consists of two Sub-Clauses. Sub-Clause (1) deals with 'Possession of Site' and is not altered. Sub-Clause (2) deals with 'Wayleaves etc.' and is altered to Sub-Clause (3). This is to allow the insertion of a new Sub-Clause (2); this new Sub-Clause has two possible forms relating to contracts involving the use of the land which will ultimately be a motorway service area.

---

*Increases are awarded automatically.

Clause 52(3) is altered unless there is no dayworks schedule included in the contract. (The guidance notes adjacent to the 'Dayworks' section found later in the Model Document suggest that a dayworks schedule should be included if there is a 'reasonable' amount of daywork). There is no definition of what is 'reasonable'. The purpose is to effect a change in the standard *Schedules of Dayworks Carried Out Incidental To Contract Work*[1] published by the Federation of Civil Engineering Contractors where it is deployed in the contract. In the latter the 'Amount of Wages' is defined as the amount which would accord with the 'Working Rule Agreement of the Civil Engineering Construction Conciliation Board for Great Britain ... or such other rule' etc. This change alters the definition of the 'Amount of Wages' to the sums paid to the work force.

Clause 57 of the *5th Edition* defines the method of measurement as the 'Civil Engineering Method of Measurement'. The Model Contract alters it to the 'Method of Measurement for Highway Works' in line with the 'Preambles to Bill of Quantities'.

Clause 60(2) is amended to empower the Engineer to issue an interim certificate of payment less than the minimum value stated in the Appendix to the Form of Tender after the Certificate of Completion has been issued.

Clause 60(6) is amended so that interest on overdue payments is paid at the rate of 1 per cent above specified Base Lending Rates as opposed to the 2 per cent stated in the *5th Edition of the ICE Conditions of Contract*.

Clause 66 is deleted and replaced with an amended version.

Sub-Clause (1) contains similar provisions to those that are in Sub-Clauses (1), (2), (3)(b), (4)(a), (4)(b) and (5)(a) of the Clause in the *5th Edition*. There are some differences. In this version the referral of a dispute or difference between the Employer and the Contractor to the Engineer for a decision need not be in writing. There is no requirement for the Employer to give effect to the Engineer's decision. The main difference, however, is contained in Sub-Clause (2) and it is the limitations of when arbitration can be sought—except where the dispute relates to

(a) Clause 12 ('Adverse Physical Conditions or Artificial Obstructions'), or
(b) the Engineer withholding any certificate, or
(c) the withholding of any portion of retention under Clause 60 ('Payment of Retention Money'), or
(d) the exercise of the Engineer's power to give a certificate under Clause 63(1) ('Forfeiture of the Contract')

then reference to arbitration cannot take place until after completion or alleged completion of the Works.

Sub-Clause (3) allows the Vice-President to act where the President is unable so to do.

The version of Clause 66 which is contained in the Model Contract Document is the version which appeared in the *5th Edition of the ICE Conditions of Contract* prior to its revision in 1979.

There then follows a series of new Clauses.

Clause 74 allows the deduction of any sums due by the Contractor to be deducted from the amounts due to him under the contract including amounts due to another Department or Office of Her Majesty's Government. In other words the right of set-off against unrelated alleged debts.

Clause 76 deals with railway track possessions with the Contractor being liable for any expenses which are fixed by the Engineer but are not, in the Contractor's opinion, necessary for compliance with his obligations under the contract. If the Contractor is denied such costs then he may have a claim under Clause 5 ('Instructions Related to Ambiguities or Discrepancies'), Clause 7 ('Delay in Issue of Further Drawings, Specifications or Instructions'), Clause 12 ('Adverse Physical Conditions or Artificial Obstructions'), Clause 13 ('Delay or Disruption Due to the Engineer Issuing Instructions or Directions'), Clause 42 ('Possession of Site'), Clause 44 ('Extension of Time'), Clause 51 ('Variations'), Clause 52 ('Valuation of Variations') or Clause 56 ('Increase or Decrease of Rate') all, of course, depending on the circumstances.

Clause 77 sets out the details of the contract price fluctuations provisions in relation to civil engineering Works and this replaces the equivalent provision which is contained in the *5th Edition* as an insert. This makes allowance for higher (or lower) payment in line with changes in the cost of labour, plant and materials.

Clause 78 is the contract price fluctuations provisions in relation to fabricated structural steelwork.

Clause 79 is the link Clause which deals with the situation where a contract contains a significant amount of both civil engineering Works and fabricated structural steelwork.

Clause 80 completes revisions to the *Conditions of Contract* and is related to the prohibition of giving information to the media without the written permission of the Engineer.

There then follows a large number of special requirements in relation to bodies whose plant or equipment may be affected by the Works, e.g. Electricity Companies, British Gas etc.

Thereafter the Model dictates that the 'Preamble to the Specification' and the 'Specification for Highway Works Schedule of Pages and Relevant Publication Dates' should follow. Both of these documents can be found in Series NG 000 Introduction of Volume 2: 'Notes for Guidance on the Specification for Highway Works' and are discussed in detail later. The Model goes on to state that both the Preamble and the Schedule should be reproduced unaltered and bound in the Specification with the numbered Appendices. (This requirement is repeated in Series NG 000 Introduction, Volume 2. The numbered Appendices are also discussed in detail later.

The next part of the Model dictates that the 'Preambles to the Bill of Quantities' and the 'Method of Measurement for Highway Works Schedule of Pages and Relevant Publication Dates' should follow. Amendments to Chapter IV should be inserted immediately after the preambles i.e. rogue or non-standard items listing the 'Units', 'Measurement', 'Itemisation' and 'Item Coverage' for any new 'Marginal Heading', (see the section on Volume 4: 'Bills of Quantities for Highway Works' later for more on this important topic).

The Model then states that the bill of quantities should follow. The guidance notes (on the left hand page) state that the bill of quantities *must* be prepared in accordance with the 'Method of Measurement for Highway Works' and mentions the 1991 Edition. It is suggested that the 1991 Edition as amended by the August 1993 and August 1994 Editions is what is actually meant. Volume 0, Section 2, Part 5, SD6/95 (discussed later) deals with the 1994 amendments and it is not particularly clear as it does not specifically mention the amendments to Volume 4 in the text but lists the alteration thereto in a table at the rear of the document.

So, the bill of quantities consists of

(*a*) the 'Preambles to Bill of Quantities', followed by
(*b*) the 'Method of Measurement for Highway Works Schedule of Pages and Relevant Publication Dates', followed by
(*c*) any amendments to Chapter IV of the 'Method of Measurement for Highway Works', and finally
(*d*) the bill of quantities itself.

The 'Preambles to Bill of Quantities' and the 'Method of Measurement for Highway Works Schedule of Pages and Relevant Publication Dates' are given in Volume 4: 'Bills of Quantities for Highway Works', Section 1, Chapter III. It states that the Schedule is to be 'reproduced unaltered and bound in the Bill of Quantities with the Preambles' but does not say the same about the Preambles. (It does however say that both the Preambles and the Schedule should be reproduced unaltered and bound in the bill of quantities in Volume 4, Section 1, Chapter III.) Any additions or amendments which are not covered by the existing 'Method of Measurement for Highway Works' have to be prepared in the same format as the existing items. No contingencies item is to be included.

The Model then deals with the subject of dayworks. The guidance notes adjacent to the dayworks section suggest that a dayworks schedule should be included if there is a 'reasonable' amount of daywork but the word is not defined and merely states that it will 'depend on the variables inherent in the individual contract' which is completely meaningless. Where there is no schedule included the Federation Schedule[3] is to be used. The guidance notes register the fact that the Highways Agency is 'increasingly concerned that a small but rapidly growing proportion of work which should more correctly be quantified and itemised within the Bill of Quantities is being set against Daywork by some of those who prepare tender documents for its Works'.

This provision reflects an increasing dismay with dayworks amongst certain Employers. The guidance notes talk about 'written justification' for amounts of dayworks and thereafter 'scrutiny by auditors' in one sentence. Dayworks are related to Clause 52(3) of the *5th Edition of the ICE Conditions of Contract* which, among other issues, imports the *Schedule of Dayworks Carried Out Incidental To Contract Work*[3] produced by the Federation of Civil Engineering Contractors as the basis for payment. The amount of dayworks in any tender is a function of the 'variables inherent' in the contract but the maximum is set at banded percentages of the value of the contract work. The schedule specifies a percentage which is added to the amount of wages to cover statutory and other charges including

(*a*) national insurance
(*b*) normal contract works, third party and employer's liability insurances
(*c*) annual and public holidays with pay and benefit scheme
(*d*) industrial training levy
(*e*) redundancy payments contribution
(*f*) Contracts of Employment Act
(*g*) site supervision and staff
(*h*) small tools
(*i*) protective clothing
(*j*) head offices charges and profit.

The Schedule sets this percentage and, in the final edition, it is 148 per cent. (The Federation of Civil Engineering Contractors ceased to exist on 15 November 1996.)

Contrast this with the approach in the Model Contract which adds a 'percentage adjustment' to the basic cost of labour. This adjustment is to include statutory charges and

(a) all other amounts of every kind paid in accordance with the rules of the appropriate wage fixing body not included in the basic cost of labour
(b) non-productive overtime and wages
(c) all bonuses, payments and incentives in excess of the basic cost of labour
(d) national insurance and surcharge
(e) Works, third party and employers liability insurances
(f) annual and public holidays with pay and benefit scheme
(g) non-contributory sick pay scheme
(h) industrial training levy
(i) redundancy payments contributions
(j) Contracts of Employment Act
(k) superintendence and supervision
(l) small tools
(m) protective clothing
(n) transport to and from site provided by the Contractor
(o) transport within the site
(p) welfare facilities
(q) use of sub-contractors
(r) establishment charges, overheads and profit.

The Highways Agency has clearly learned from its experiences by including a large number of items which previously were not to be found in the list. However, it is suggested that one item is unfair and contrasts with a fundamental element of the approach used in the preparation of the contract as dictated by the *Manual of Contract Documents for Highway Works* itself. The inclusion of non-productive overtime represents an element which cannot be predicted by the Contractor. The timing of dayworks operations may well be due to some default of the Engineer. It may come about as a result of some matter which is outwith the control of all parties. Does the Contractor carry it out during normal working hours in the hope that the Engineer will agree to an extension of time? Is he in breach of the common law duty to minimise costs by not working outside of normal working hours because his guess about the proportion of non-productive overtime was nil or insufficient? It would be far more equitable to pay overtime rates where worked. After all, daywork operations are instructed entirely at the whim of the Engineer. It is suggested that the Engineer would have some difficulty in repudiating a claim for non-productive overtime which arises from dayworks carried out by the Contractor in circumstances which he could not reasonably be expected to foresee.

The approach to plant elements is also changed under the Model Contract Document. Tenderers have to insert a percentage adjustment to be added to the charges for plant. Charges are the hire rates from the Federation Schedule.[3] The Model requires the tenderer to have made allowance for the following which would not be included under the Federation schedule.

(a) plant which is hired (under the Federation Schedule hired plant is charged at the invoice value plus $12\frac{1}{2}$ per cent) or leased
(b) plant which is made available for use but not used is to be charged at two-thirds the rate in the schedule subject to a maximum of eight hours per day or 24 hours per day
(c) fuel distribution
(d) transporting to and from the site.

Payment for materials in dayworks operations is again different from the approach used in the Federation schedule[3] with the Model attempting to anticipate every source of cost. Again a percentage adjustment is to be applied to the cost of materials with the figure being inserted by the tenderers.

There then follows a summary where the tenderers are to insert the percentage adjustments for firstly labour then plant and finally materials. It generously precludes negative percentages in both the labour and materials sections.

Apart from a note on the removal of the painting index, the dayworks section concludes the Model Contract Document.

*Part 3. Model Contract Document for Highway Works—Scotland*
Those matters discussed in the previous section, which deals with the Model Contract for England, are the same in the version for Scotland unless noted hereunder. The version of Part 3 which was supplied by HMSO until at least late 1996 is dated November 1992 (it is a reprint of what was called Annex B to SD5) and contains provisions for the use of the 6th Edition of the *Manual of Contract Documents for Highway Works* as well as the 7th Edition. In fact, The Scottish Office now uses an amendment which is vastly different from that which is published by HMSO. The version published by HMSO is not discussed here since it is most unlikely to find any substantial use in future.

The following relates to the Model Contract currently (June 1997) supplied by The Scottish Office.

Like its equivalent for England, Part 3 is divided into two sections. The first section deals with the 'Instructions for Tendering' whilst the second section is the Model Contract itself. The document is very similar to the version for England and the only change worthy of note is that the new Clause 76 ('Railway Possessions') is not included.

The *Design Manual for Roads and Bridges* does introduce some changes to the provisions of the *5th Edition of the ICE Conditions of Contract* for lane rental contracts in Scotland. Various Clauses are amended or deleted and details are given in Technical Memorandum SH4/91 which can be found in Section 1 of Volume 5 of the *Design Manual for Roads and Bridges*.

*Part 4. Model Contract Document for Highway Works—Wales* Those matters discussed in the section headed Part 2, which deals with the Model Contract for England, are the same in the version for Wales unless noted hereunder.

Like its equivalent for England, Part 4 is divided into two sections: the first section deals with the 'Instructions for Tendering'; the second section is the Model Contract itself.

*Instructions for Tendering.* Although, in some cases they are set out differently, the content is essentially as Part 2 discussed above.

*The Model Contract.* Clause 14 is amended to require the Contractor to supply to the Engineer in such detail as the latter requires the make-up of his tender showing for each item the allowance for labour, plant, equipment, materials, overheads and profit. This is quite a good idea from the standpoint of the Employer as it can be used to evaluate any extra payments without the information being presented in a manner which maximises his benefit.

*Part 5. Model Contract Document for Highway Works—Northern Ireland* Those matters discussed in the previous section, which deals with the Model Contract for England, are the same in the version for Northern Ireland unless noted hereunder. The version of Part 5 which was supplied by HMSO until at least June 1997 is dated November 1992 (it is a reprint of what was called Annex B to SD5) and contains provisions for the use of the 6th Edition of the *Manual of Contract Documents for Highway Works* as well as the 7th Edition. In fact, The Department of the Environment for Northern Ireland now uses an amendment which is vastly different from that which is published by HMSO. The version published by HMSO is not discussed here since it is most unlikely to find any substantial use in future.

The following relates to the Model Contract currently (March 1997) supplied by The Department of the Environment for Northern Ireland.

Like its equivalent for England, Part 5 is divided into two sections. The first section deals with the 'Instructions for Tendering' whilst the second section is the Model Contract itself. The document is very similar to the version for England.

Clause 27 which relates to the Public Utilities Street Works Act 1950/the New Roads and Street Works Act 1991 has been deleted.

Clause 42 is not amended.

Clause 52(3) is not amended and thus the *Schedule of Dayworks*[3] applies without alteration.

Clause 56 is amended to permit the use of a balancing item in tenders.

Clause 60(4) is amended to increase retention to 10 per cent of the amount due to the Contractor.

Clause 66 is changed to render the contract subject to the law of Northern Ireland.

Clause 77 is to be used where the contract period is less than two years and renders the contract fixed price.

Clauses 78, 79 and 80 are the price fluctuations clauses and are the equivalent of Clauses 77, 78 and 79 in the version for England.

Clause 81 is that which deals with privacy of information relating to the contract and is the same as Clause 80 in the conditions for England.

Clause 82 requires approval of any security firms used in the contract.

Clause 83 repeats the provision contained in Clause 66 that makes the law of Northern Ireland applicable to the contract.

### 3.3.3. Volume 0, Section 2: Implementing Standards

*Part 1. Implementation of Specification for Highway Works and Notes for Guidance*  This document dated July 1992 relates to the 1991 Edition of the *Manual of Contract Documents for Highway Works*. It lists the standard 27 Series (Series 000 to Series 2600 inclusive) and the Lettered Appendices (Appendix A to Appendix G inclusive). It explains that the Notes for Guidance for Highway Works gives details of the compilation of Numbered Appendices. It states, and this is important, that the 'documents contained in Volumes 0, 3 and 4 of the Manual of Contract Documents for Highway Works shall be used when preparing contracts incorporating the SHW'.

Inexplicably it does not mention Volume 1: 'Specification for Highway Works' and Volume 2: 'Notes for Guidance on the Specification for Highway Works' but it does import both under the heading 'Implementation' (see Part 2 following).

Volume 0 is entitled 'Model Contract Document for Major Works and Implementation Requirements'; Volume 1 is entitled the 'Specification for Highway Works'; Volume 2 is entitled 'Notes for Guidance on the Specification for Highway Works'; Volume 3 is entitled 'Highway Construction Details'; Volume 4 is entitled 'Bills of Quantities for Highway Works'. On occasions, it may be stated, in a claims situation, that certain parts of the *Manual of Contract Documents for Highway Works* are not contractual (usually the Notes for Guidance). How can they be non-contractual when their use is mandatory?

*Part 2. Implementation of Highway Construction Details* Volume 3 contains standard details in drawing format. The interesting point is that it cites the use of Volumes 0 to 2 and Volume 4 (see above).

*Part 3. Preparation of Bill of Quantities for Highway Works* This document requires the use of Volumes 0 to 3 of the *Manual of Contract Documents for Highway Works* when preparing contracts incorporating Volume 4. It also includes a standard layout for bills of quantities.

*Part 5. Implementation of 1994 Annual Amendments to Specification for Highway Works and Notes for Guidance, Highway Construction Details and Preparation of Bill of Quantities for Highway Works* This document relates to the August 1994 amendments to the December 1991 edition of the *Manual of Contract Documents for Highway Works.* It contains advice on how the August 1994 documents are to be incorporated into contracts, the date of implementation etc. It also contains schedules of amendments to Volumes 1, 2, 3 and 4. It does not clearly refer to the amended versions of Volume 3: 'Highway Construction Details' and Volume 4: 'Bills of Quantities for Highway Works' in the text but they are listed at the end of this document.

## 3.4. Volume 1: 'Specification for Highway Works'

### 3.4.1. General

This Volume dictates the methods by which the Works are to be carried out. It contains a very wide range of requirements which extend from, for example, specifying the type of reflective jackets to be worn by all staff on motorways to detailed requirements for earthworks materials. It is a combination of what are termed 'method' specifications and 'end product' specifications (also called results specifications). A method specification describes the way in which some aspect of construction is executed whereas an end product specification defines one or more parameters which must be met when the work has been completed. Many items are a combination of both types of specification e.g. bituminous mixtures where there are method specification requirements related to compaction amongst other matters and end product specification requirements related to the finished level. Volume 1: 'Specification for Highway Works' consists of 27 sections numbered Series 000, Series 100, Series 00 and so on up to Series 2600 and 7 Lettered Appendices designated Appendix A to Appendix G. Each Series addresses a different area of activity and the Lettered Appendices contain standard information and requirements. Within each Series, the Clauses are numbered, for example 801, 802 etc. (except for Series 000 which starts at Clause 000 then 001 etc.).

The titles of the Series and Appendices are set out below.

| | |
|---|---|
| Series 000 | Introduction |
| Series 100 | Preliminaries |
| Series 200 | Site Clearance |
| Series 300 | Fencing and Environmental Barriers |
| Series 400 | Safety Fences, Safety Barriers and Pedestrian Guardrails |
| Series 500 | Drainage and Service Ducts |
| Series 600 | Earthworks |
| Series 700 | Road Pavements—General |
| Series 800 | Road Pavements—Unbound Materials |
| Series 900 | Road Pavements—Bituminous Bound Materials |
| Series 1000 | Road Pavements—Concrete and Cement Bound Materials |
| Series 1100 | Kerbs, Footways and Paved Areas |
| Series 1200 | Traffic Signs |
| Series 1300 | Road Lighting Columns and Brackets |
| Series 1400 | Electrical Work for Road Lighting and Traffic Signs |
| Series 1500 | Motorway Communications |

# 158 | CLAIMS ON HIGHWAY CONTRACTS

| | |
|---|---|
| Series 1600 | Piling and Diaphragm Walling |
| Series 1700 | Structural Concrete |
| Series 1800 | Structural Steelwork |
| Series 1900 | Protection of Steelwork Against Corrosion |
| Series 2000 | Waterproofing for Concrete Structures |
| Series 2100 | Bridge Bearings |
| Series 2200 | Parapets |
| Series 2300 | Bridge Expansion Joints and Sealing of Gaps |
| Series 2400 | Bridgework, Blockwork and Stonework |
| Series 2500 | Special Structures |
| Series 2600 | Miscellaneous |
| | |
| Appendix A | Quality Management Schemes |
| Appendix B | Product Certification Schemes |
| Appendix C | British Board of Agrément Roads and Bridges Certificates |
| Appendix D | Statutory Type Approval |
| Appendix E | Departmental Type Approval/Registration |
| Appendix F | Publications Referred to in the Specification |
| Appendix G | Petrographical Examination of Aggregates for Alkali—Silica Reaction |

Series 000 of Volume 1 is entitled 'Introduction' and covers various matters. Like all the introductory series in the *Manual of Contract Documents for Highway Works*, it is well worth studying in detail particularly to those interested in the commercial aspects of contracts. Clause 114 of Series 100 which relates to monthly statements is also worthy of consideration.

*3.4.2. Series 000—Introduction*

*Sub-Clause 000.1* There are 27 numbered Series (including the Introduction, Series 000) and 7 Lettered Appendices in the Specification. The 'Schedule of Pages and Relevant Publication Dates' specifies the versions which apply to any contract.

*Sub-Clause 001.1* This relates to national alterations of the overseeing departments of Scotland, Wales and Northern Ireland. It states that additional or substitute specification requirements of these departments (identified by the use of the suffix SO, WO or NI) are at the end of each Series or Lettered Appendix. Note that these requirements are deemed to be included in contracts administered by these departments unless stated otherwise in Appendix 0/5. A sample of Appendix 0/5 is included at the end of Series NG 000 of Volume 2: 'Notes for Guidance on the Specification for Highway Works'. Clauses with suffix SO, WO or NI are deemed to replace Clauses having the same number but without the suffix in contracts overseen by these departments. This is an important point.

*Sub-Clause 002.1* Unless expressly stated otherwise, BS 6100, Glossary of Building and Civil Engineering Terms definitions apply to the Specification and associated documents. It can be argued with some confidence that this applies beyond the documents included in the *Manual of Contract Documents for Highway Works.*

*Sub-Clause 002.2* Abbreviations are those given in BS 5775, Specification for Quantities, Units and Symbols (Parts 0 to 13) along with those listed in Table 0/1 which can be found within this Series of Volume 1: 'Specification for Highway Works'.

*Sub-Clause 003.1* Appendices are always termed either Lettered Appendices or Numbered Appendices. Lettered Appendices contain standard information and requirements while Numbered Appendices contain contract-specific information.

*Sub-Clause 003.2* Numbered Appendices contain contract-specific information and requirements and those which are incorporated in the contract are listed in Appendix 0/3, an example of which can be found at the end of Series NG 000 of Volume 2: 'Notes for Guidance on the Specification for Highway Works'.

*Sub-Clause 004.1* The British Standards (and Drafts for Development), Department of Transport publications, Transport Research Laboratory Reports, International Standards, Acts and Statutory Instruments, Quality Assurance Certification Bodies and other publications are listed in Appendix F at the end of Volume 1: 'Specification for Highway Works'.

*Sub-Clause 004.2* The dates of British Standards and British Standard Codes of Practice which are listed in the contract but do not have a date included are those which were current at the reference date which should be stated in the Appendix to the Form of Tender.

*Sub-Clause 004.3* The dates of all other references are as stated in Appendix F or, if not stated, then the reference date. Note that it includes all amendments published up to that date. Amendments can contain important information which will often have a commercial significance.

*Sub-Clause 004.4*   Harmonised European Standards and European Standards which supersede a British Standard and are issued prior to the reference date take precedence over the British Standard.

*Sub-Clause 005.1*   Thicknesses of materials are the finished or compacted thicknesses.

*Sub-Clause 005.2*   Tolerances are those given in the Specification, the drawings and the publications listed in Appendix F.

### 3.4.3.   Series 100—Preliminaries

Sub-Clause 114 'Monthly Statements' requires the submission of monthly statements as required by Clause 60(1) of the *5th Edition of the ICE Conditions of Contract* to be presented in accordance with Appendix 1/14. It instructs that Statements are to be produced in a format similar to that of the bills of quantities (see Volume 4: 'Bills of Quantities for Highway Works'. In respect of all other matters to which the Contractor considers himself entitled they are to be included in the statement. The Contractor shall allow the Engineer to inspect all details related to materials on site and goods the ownership of which is vested in the Employer.

## 3.5.   Volume 2: 'Notes for Guidance on the Specification for Highway Works'

### 3.5.1.   General

Volume 2: 'Notes for Guidance on the Specification for Highway Works' is designed to provide guidance to Engineers involved in employing Volume 1: 'Specification for Highway Works' on particular contracts. It is not included in the list of documents listed in the form of agreement (see Volume 0: 'Model Contract Document for Major Works and Implementation Requirements'). Some may argue that it is non-contractual but it has been seen that it is stated in Volume 0 that it is used in producing the specification and consequently it is very difficult to see how it is anything other than contractual.

The requirements and recommendations contained in Volume 2

are often ignored or misapplied. It is often useful to examine a particular contract specification to see if it has been compiled in accordance with this document.

The section following takes a detailed look at those areas of Volume 2 which may be significant in terms of commercial matters or contractual claims.

### 3.5.2. Series NG 000—Introduction

*Sub-Clause NG 000.1* The numbered Clauses are directly related to the Clauses in Volume 1: 'Specification for Highway Works' having the same number but without the prefix NG. Not every Clause in Volume 1 has an equivalent Clause in Volume 2.

*Sub-Clause NG 000.2* The word 'Engineer' in Volume 2 means the designer/compiler.

*Sub-Clause NG 000.3* There must be a Preamble to the Specification containing a Schedule of Pages and Relevant Publication Dates which lists the date of publication of each page of the Specification. As has been seen these are to be included in the contract unaltered and bound in with the Specification.

*Sub-Clause NG 000.4* Proprietary work, goods or materials must not be used unless their functions cannot be described with sufficient precision and ease of understanding and in such circumstances the phrase 'or equivalent' must be added.

*Sub-Clause NG 000.5* The use of Volume 1: 'Specification for Highway Works' is 'mandatory' on trunk roads including trunk road motorways.

*Sub-Clause NG 000.6* The Specification as set out in Volume 1 need not be reproduced since it is imported by reference.

*Sub-Clause NG 000.7* Contract-specific requirements are contained in the Numbered Appendices and are dealt with in one of two ways

(a) specific requirements referred to in the national Specification are listed in Table NG 0/1 which can be found at the end of Section NG 000 and
(b) alternatives to national requirements are listed in Table NG 0/2 which can also be found at the end of Section NG 000.

Table NG 0/1 consists of rows of Numbered Appendices with national requirements under which is a list of the Sub-Clauses within which the Numbered Appendices are mentioned. Table NG 0/2 consists of columns of Sub-Clauses with textual descriptions which contain contract-specific requirements. These tables are not included in the main Specification and are included for use as check lists by the compiler (this fact is explained in Sub-Clause NG 003.15).

*Sub-Clause NG 000.8* Volume 1: 'Specification for Highway Works' is to be used as it stands without alteration wherever possible. Under no circumstances should replacement pages be produced to incorporate contract-specific requirements.

*Sub-Clause NG 000.9* Numbered Appendices 0/1 to 0/5 and 1/1 onwards should be bound in one volume or two as appropriate and clearly titled as to the content.

*Sub-Clause NG 002.1* Terms and abbreviations used in additional and substitute clauses and in minor alterations should be consistent with those in Sub-Clause 002 in Volume 1.

*Sub-Clause NG 002.2* Departmental standards for the Department of Transport and where possible the overseeing departments of Scotland, Wales and Northern Ireland and forming part of Volume 0: 'Model Contract Document for Major Works and Implementation Requirements' give implementing information and other instructions on the use of Volume 1: 'Specification for Highway Works' and related departmental procedures within their geographical areas. In addition the departmental standards of the overseeing departments of Scotland, Wales and Northern Ireland will contain any special national alterations to be included in Appendix 0/5.

*Sub-Clause NG 003.1* All Lettered Appendices are national requirements and should be used without alteration except for Appendix F ('Publications Referred to in the Specification').

*Sub-Clause NG 003.2* Lettered Appendices A to G contain details of accepted quality management schemes, product certification schemes, British Board of Agrément Roads and Bridges Certificates, statutory type approval, departmental type approval/registration, publications referred to in the specification and petrographical examination of aggregates for alkali–silica reaction respectively.

*Sub-Clause NG 003.3* Appendix F is a list of all publications referred to in Volume 1 : 'Specification for Highway Works'. Any alterations to Appendix F are to be listed in Appendix 0/2.

*Sub-Clause NG 003.4* Numbered Appendices are to be included in the documents. Where contract-specific information is to be included on the drawings then cross reference should be made in the Numbered Appendix to the relevant drawing number.

*Sub-Clause NG 003.5* Numbered Appendices are to be compiled using the advice in this document i.e. Volume 2: 'Notes for Guidance on the Specification for Highway Works'. Some Numbered Appendices are to be completed by the tenderer or Contractor.

*Sub-Clause NG 003.6* Numbered Appendices 1/1 onwards are to be used to extend the information in Volume 1: 'Specification for Highway Works' Clauses, Tables and Figures NOT (upper case in Volume 2: 'Notes for Guidance on the Specification for Highway Works') to change it. Where the contract requires new methods or special requirements then these are to be introduced by additional or substitute Clauses in Appendix 0/1 or by minor alterations in Appendix 0/2.

*Sub-Clause NG 003.7* New Numbered Appendices can be used to extend contract-specific alterations.

*Sub-Clause NG 003.8* The 'Zero' Series Numbered Appendices (0/1 to 0/5) are to be compiled in accordance with the examples given, see end of Series NG 000 for these examples.

*Sub-Clause NG 003.9* Contract-specific alterations to Volume 1: 'Specification for Highway Works' should be described in Appendices 0/1 and 0/2 (again, see end of Series NG 000 for examples) but should take into account

(a) that requirements relating to Clauses apply equally to Tables and Figures
(b) that care should be taken to ensure compatibility between Clauses and any Tables and Figures which are altered
(c) that text which conflicts with or duplicates the provisions of the *Conditions of Contract* should be avoided.

*Sub-Clause NG 003.10* Appendix 0/1 contains contract-specific alterations and consists of a list of additional, substitute and cancelled clauses employing the suffices AR, SR and CR respectively. If there are none then the word 'NONE' should be written in Appendix 0/1. Following the three (or one or two as appropriate) lists, the actual additional and substitute Clauses are stated.

*Sub-Clause NG 003.11* Appendix 0/2 contains contract-specific minor alterations to existing Clauses, Tables and Figures and is enacted by using the phrase 'delete and insert' or 'add new Clause'. If there are none then the word 'NONE' should be written in Appendix 0/2.

*Sub-Clause NG 003.12* Appendix 0/3 should contain a complete list of Numbered Appendices included in the contract. There are two lists, List A and List B. List A is for Numbered Appendices referred to in Volume 1: 'Specification for Highway Works' and List B is for Numbered Appendices devised by the Engineer. Where a numbered Appendix in List A is not used it should be marked 'Not Used' beside its number. Those to be completed by the tenderer must be identified and blank pro formas supplied, copied from Volume 2: 'Notes for Guidance on the Specification for Highway Works'.

*Sub-Clause NG 003.13* Appendix 0/4 should contain a complete list of drawings for the contract including those in Volume 3: 'Highway Construction Details'.

*Sub-Clause NG 003.14* Appendix 0/5 should contain special national alterations of the Overseeing Department of Scotland, Wales or Northern Ireland.

*Sub-Clause NG 003.15* Tables NG 0/1 to NG 0/3 should be used by the compiler as check lists with

(a) Table NG 0/1 identifying Sub-Clauses which contain a reference to a Numbered Appendix which will contain contract-specific requirements
(b) Table NG 0/2 identifying Sub-Clauses etc. which are complete but which recognize that contract-specific requirements may be applicable
(c) Table NG 0/3 identifying Sub-Clauses which require the Contractor to submit information to the Engineer.

*Sub-Clause NG 004.1*  The reference date establishes the date of each British Standard to be used in the contract unless a date for any is specified in Appendix F.

*Sub-Clause NG 004.2*  The text of Volume 1: 'Specification for Highway Works' and Volume 2: 'Notes for Guidance on the Specification for Highway Works' is based on the British Standards and amendments published on or before 1 March 1993. Any alterations which arise because of Harmonised European Standards will be drawn up by the overseeing departments at regular intervals and published by HMSO.

*Sub-Clause NG 004.3*  The Engineer is required to ascertain whether or not any amendments or new editions of the reference documents have been published since the last published alteration to Volume 1 and, if appropriate, include contract-specific alterations in Appendix 0/1 or Appendix 0/2.

*Sub-Clause NG 004.5*  The Engineer should consider whether reference document amendments published after the contract award should be incorporated in the contract. It states that amendments, such as those making editorial changes as published by the British Standard Institution, will often have negligible cost implications and should usually be adopted.

*Sub-Clause NG 004.6*  The list of British Standards and other publications is given in Annex A which is contained at the end of Volume 2: 'Notes for Guidance on the Specification for Highway Works'.

*Sub-Clause NG 005.1*  Unnecessarily fine tolerances have the effect of increasing prices and should be avoided.

*Tables NG 0/1, 0/2 and 0/3*  These check list Tables, discussed in NG 003.15, are printed here.

*Preamble to the Specification and Schedule of Pages and Relevant Publication Dates*  The 'Preamble' and the 'Schedule' *must* be reproduced unaltered and bound in the Specification along with the Numbered Appendices (see the note at the end of the 'Standard Preamble' and also Volume 0: 'Model Contract Document for Major

Works and Implementation Requirements' in the section after the special requirements for the utilities). The Schedule is a list of the date of publication of all pages of Volume 1: 'Specification for Highway Works'.

Suffixes to Clauses can be A, AR, S, SR, C or CR, with A meaning additional, S meaning substitute and C meaning cancelled. The R indicates that the change is contract-specific; if R is not used then the amendment originates from the overseeing department of Scotland, Wales or Northern Ireland. All should be listed in Appendix 0/5.

The Numbered Appendices always takes precedence over any provision of Volume 1: 'Specification for Highway Works'. Additionally Appendices 0/1 and 0/2 take precedence over Appendix 0/5.

In altered Clauses, original Tables or Figures apply unless they are amended in the altered Clause.

Where a Clause in Volume 1: 'Specification for Highway Works' relates to work, goods or materials which are not required in the contract it shall be deemed not to apply.

Superscript$^{8/93}$ means the August 1993 version.

The use of the hash (#) indicates an alteration by the overseeing department of Scotland, Wales or Northern Ireland. They can only be used within the countries to which they apply. They replace the corresponding Clauses in the main text or are additional thereto as appropriate. The substitute Clauses are located at the end of the relevant Series together with the National Clauses of the overseeing departments.

The August 1994 Volume 1: 'Specification for Highway Works' is to be used in conjunction with the August 1994 Volume 4: 'Bills of Quantities for Highway Works' and the February 1994 versions of the Model Contract Document for Highway Works (Section 1 of Volume 0: 'Model Contract Document for Major Works and Implementation Requirements').

### 3.5.3. Series NG 100—Preliminaries

*Sub-Clause NG 105.2* Those tests which are to be undertaken by the Contractor along with the test certificates to be supplied should be taken from Table NG 1/1 which is included at the end of this Series as there are no items which cover provision in Volume 4: 'Bills of Quantities for Highway Works'. They should then be listed in Appendix 1/5 and again there is a sample at the end of this Series. Some of the frequencies required in this Appendix are unreasonably high. There is a note at the end of the Table NG1/1—

Typical Testing Details, which is contained in Series 100 of Volume 2: 'Notes for Guidance on the Specification for Highway Works' and suggests that these are for guidance only and that where material properties are consistently well above minimum or well below maximum requirements then the frequency can be reduced. The difficulty is that this statement is not reproduced in the contract although it can be argued that this statement is quasi-contractual and its advice is often ignored and so, for example, a frost heave certificate (Clause 705) is required every 500 m$^3$ when, in fact, tests of this nature should be on a one per source basis. This is particularly true when the *Manual of Contract Documents for Highway Works* is used for non-trunk road contracts—a very common practice.

*Sub-Clause NG 105.3*   Other required testing not covered in Table NG 1/1 but required in contract-specific Clauses should be scheduled in Appendix 1/5 or 1/6 as appropriate (samples of these are provided at the end of this Series).

*Sub-Clause NG 105.5*   Table NG 1/1 lists frequencies against specific tests. These are indicative only and the Engineer should decide the frequency taking into account various factors. Where a British Standard is listed or Specification Clause number is mentioned, the frequency is specified therein and should not normally be changed.

*Sub-Clause NG 105.6*   Details of the provision and delivery of samples should be included in Appendix 1/6 as there are no items which cover same in Volume 4: 'Bills of Quantities for Highway Works'.

## 3.6.   Volume 4: 'Bills of Quantities for Highway Works'

### 3.6.1.   General

This Volume is probably the most important part of the *Manual of Contract Documents for Highway Works* for those who are involved in the commercial aspects of a contract. It consists of three sections

- Section 1: Method of Measurement for Highway Works
- Section 2: Notes for Guidance on the Method of Measurement for Highway Works
- Section 3: Library of Standard Item Descriptions for Highway Works.

Like Volume 1: 'Specification for Highway Works', each of the above sections is split up into Series which is similar to the numbering format of the specification with a number of differences. There is no Series 000 Introduction, there is a Series 2700 Accommodation Works and Works for Statutory Undertakers and Series 800, 900, 1000 and 2600 are not taken up. Series 800, 900 and 100 are not taken up since the items therein are dealt with under Series 700. Series 2600 is not taken up since it consists of general materials such as mortar, concrete for ancillary purposes and thermoplastic node markers which are not items in their own right. The list of Series in Section 1 is as follows.

|  |  |
| --- | --- |
| Series 100 | Preliminaries |
| Series 200 | Site Clearance |
| Series 300 | Fencing and Environmental Barriers |
| Series 400 | Safety Fences, Safety Barriers and Pedestrian Guard-rails |
| Series 500 | Drainage and Service Ducts |
| Series 600 | Earthworks |
| Series 700 | Pavements |
| Series 800 | Not taken up |
| Series 900 | Not taken up |
| Series 1000 | Not taken up |
| Series 1100 | Kerbs, Footways and Paved Areas |
| Series 1200 | Traffic Signs |
| Series 1300 | Road Lighting Columns and Brackets |
| Series 1400 | Electrical Work for Road Lighting and Traffic Signs |
| Series 1500 | Motorway Communications |
| Series 1600 | Piling and Diaphragm Walling |
| Series 1700 | Structural Concrete |
| Series 1800 | Structural Steelwork |
| Series 1900 | Protection of Steelwork Against Corrosion |
| Series 2000 | Waterproofing for Concrete Structures |
| Series 2100 | Bridge Bearings |
| Series 2200 | Parapets |
| Series 2300 | Bridge Expansion Joints and Sealing of Gaps |
| Series 2400 | Bridgework, Blockwork and Stonework |
| Series 2500 | Special Structures |
| Series 2600 | Not taken up |
| Series 2700 | Accommodation Works and Works for Statutory Undertakers |

The list in Section 2 is the same as above but Series 800, 900, 1000 and 2600 are missed out (they are not even given a blank page as they were in Section 1). Section 3 has the Series as in Section 1 but

Series 2700 is 'Not taken up' and the following is added: Series 2800 Provisional Sums and Prime Cost Items.

This chapter looks at the important parts of this Volume concentrating on those which are likely to come into play when claims are involved.

### 3.6.2. Section 1—Method of Measurement for Highway Works

There are four chapters in the Method of Measurement section and the contents of each is considered below.

*Chapter I—Definitions* Various terms are defined.

*Chapter II—General Principles* These principles ought to be second nature to anyone pursuing claims. The method of measurement is intended to be used in conjunction with the *5th Edition of the ICE Conditions of Contract*. As discussed earlier, there seems no obvious reason why the 6th Edition has not been adopted.

The method of measurement is based on Volume 1: 'Specification for Highway Works' and Volume 3: 'Highway Construction Details' of the *Manual of Contract Documents for Highway Works* on the very important principle that full details of construction are contained in the contract. Additions or amendments to Volumes 1 or 3 which are not adequately covered by the method of measurement require amendment to suit (see Chapter III following). This is often not done or is badly done.

The work covered by each item in the bill of quantities is identified by three elements

(*a*) the Sub-Heading
(*b*) the Item Description itself

in conjunction with

(*c*) the Item Coverage for the appropriate Marginal Headings.

Examples of this are given later.

The nature and extent of the work to be performed is to be ascertained by reference to the drawings, specification and conditions of contract. This phrase is often quoted in claims negotiations and represents an important principle.

Each item description is to be consistent with and be compounded from one or more of the groups listed under the marginal headings 'Itemisation' (see Chapter IV later).

The Itemisation Tables are very important. They are divided into Groups and each Group is divided into a number of Features. However, there may be only one Group and there may be only one Feature within any Group. All items in the bill of quantities consist of an Item Description which is compiled from the Itemisation Tables. The Table may have several Groups but the item may take only one feature from one or more of the Groups. The Item Description should use a feature from as many Groups as necessary to 'identify the work required'. More than one Feature from a single Group is not permitted. Unless expressly stated otherwise in the contract, the bill of quantities is to be compounded in accordance with these requirements.

*Chapter III—Preparation of Bill of Quantities* Table 1 sets down the separate levels of identification into which the bill of quantities is to be divided.

Quantities are to be whole numbers unless the unit is tonne or hectare in which case it shall be three decimal places.

The most important elements of this chapter are set out at the end under the heading 'Preambles to Bill of Quantities'. It consists of 16 sections and it must always be included unaltered as the preamble to the bill of quantities.

*Preambles to Bill of Quantities* This section contractually confirms many of the principles employed to produce the bill of quantities, defines what is included in all items and gives guidance to the Contractor on the contents of the tender and how it should be priced. The first two sections fall under the marginal heading 'General Directions'.

The first paragraph confirms that the bill of quantities has been prepared in accordance with Section 1 (Method of Measurement for Highway Works) of Volume 4: 'Bills of Quantities for Highway Works'.

Sub-Clause 2 is very significant. It confirms the roles and interrelationship of the Sub-Headings, Item Descriptions and Item Coverages. Here, it contractually states that the 'nature and extent of the work is to be ascertained by reference to the Drawings, Specification and Conditions of Contract'. It also contains a list of matters which have to be included in the rates inserted in the tender unless expressly stated otherwise. These are known as the 'general directions' and include 'labour and costs in connection therewith'. Cost in Clause 1(5) of the *5th Edition of the ICE Conditions of Contract* 'shall be deemed to include overhead costs whether on or off the Site unless expressly stated otherwise'. 'Materials, goods storage and costs in connection therewith' and 'plant and costs in connection therewith' are also included along with various other matters.

It is worth looking at the inter-relationship of all the elements which combine to describe an item in detail as described in Sub-Clause 2 of the preambles. The work covered by any item is identified by

(a) the items listed in General Directions
(b) the Sub-Heading
(c) the Item Description
(d) the Item Coverage.

It does not matter whether the Item Coverage is a standard item or one fabricated for a non-standard item (assuming that it is compiled properly); if it is a non-standard item and one or more elements in the above are missing, then the work covered is identified by those elements which do exist and what the objective observer would conclude is covered. Often it is the case that the lack of, say, an Item Coverage makes little difference to what can be reasonably perceived to be the work covered by any item. Notwithstanding this, the Engineer is simply creating difficulties for himself (and the Contractor) by not complying with the philosophy behind the Method of Measurement.

Section 17 relates to national alterations. Where the 'hash' symbol (#) is used then that indicates that a paragraph has an alternative for Scotland, Wales or Northern Ireland. They have been produced to correspond to changes in the 'Specification for Highway Works' for those countries. The alternative replaces the original in contracts within that country. It cannot be used elsewhere.

Finally there is the obligatory Schedule of Pages and Relevant Publication Dates which must be reproduced unaltered and bound in the bill of quantities.

*Chapter IV—Units and Methods of Measurement* This chapter employs the standard 26 Series used in the specification (but without Series 000: Introduction) i.e. Series 100: Preliminaries, Series 200: Site Clearance etc. although some are not used—Series 800, 900, 1000 and 2600 and the additional Series 2700: Accommodation Works and Works for Statutory Undertakers.

This chapter consists of the 'unit of measurement' and what are termed the methods of measurement for a given sub-heading. Each method of measurement consists of an Itemisation Table followed by one or more lists of Item Coverages.

For example the method of measurement under the sub-heading 'Temporary Accommodation' gives the unit of measurement for the erection, servicing and dismantling of temporary accommodation as ..... item. The Itemisation Table is then given followed by separate item coverages for erection, servicing and dismantling items. It is not always the case that there are separate item coverages for the group 1 features.

Looking at the Itemisation Table the wording of items can be discerned. Bear in mind that a Feature from one or more Groups can be used in any item. The number used is determined only by the need for the Item Description to be adequate to describe the work covered (in conjunction with the Sub-Heading, the Item Coverage and the standard Item Coverage in the General Directions). Remember also that only one Feature from any single Group may be used and never more than one. Examination of the Itemisation Table shows what options are available and also what a straightforward process this is. On occasions, the Itemisation Table is appended by a note, the meaning of which is usually self-evident.

The Item Coverages are always preceded by the words 'shall in accordance with the Preambles to the Bill of Quantities General Directions include for:'. Thus the standard inclusions listed earlier i.e. the labour and costs in connection therewith and the supply of materials, goods, storage etc. are imported into all items in the bill of quantities unless expressly stated otherwise.

### 3.6.3. Section 2—Notes for Guidance on the Method of Measurement for Highway Works

This short section sets out basic advice for the Engineer in preparing the bill of quantities and, logically, employs the standard Series system except that Series 800, 900, 1000 and 2600 are omitted and Series 2700 Accommodation Works and Works for Statutory Undertakers is added.

*General* It is worth reading the paragraph headed 'Item Coverage' at the start of this section. It confirms the importance of the adequacy of the Item Coverages. It should be noted that the existence of an element within an Item Coverage does not mean that the Contractor has made allowance for that element. It means that the Contractor is deemed to have made allowance only if some other part of the contract documentation required it. For example, the Item Coverage for tack coat Series 700, Paragraph 24 includes '(i) admixtures and additives' but where the specification requires a standard tack coat which contains no admixtures or additives then the Contractor could not reasonably be expected to allow for their inclusion. Some people have a substantial difficulty with this proposition.

There is then the rather strange statement that coverages related to the base Item Description are not normally included, e.g. 'cement in concrete' and similarly 'those contingently and dispensably necessary to enable the work item to be completed satisfactorily, for example nuts and bolts in safety fences' but since 'the supply of materials, goods ... and costs in connection therewith' is included under the General Directions it is difficult to see the need for this statement.

'Similarly general obligations are not separately covered, e.g. obligations set out in the Conditions of Contract or covered in the Preambles to Bill of Quantities'. This is a very important principle but it could be made clearer by clarifying the phrase 'General obligations, liabilities and risks involved in the execution of the Works set forth or reasonably implied in the documents on which the tender is based' which is item (vii) in the General Directions Item Coverage.

It is confirmed that the Item Coverages closely match Volume 1: 'Specification for Highway Works' and Volume 3: 'Highway Construction Details'. Therefore, if changes are made to the Specification then the Item Coverages may have to be revised to reflect these alterations. It advises that changes to the Specification should not be made on the drawings but revised drawings may reflect changes to the Specification in which case reference to the drawings should be made in the Specification. Also Item Coverages should not be extended to include work elements which do not feature in Volume 1 or Volume 3.

Note that some Item Coverages are cross referenced to others and therefore include those Item Coverages e.g. the Item Coverage for drains, sewers etc. also mentions and thus imports Series 600, paragraphs 17 and 18 which are the Item Coverages for excavation of two different grades of acceptable material.

Extra over items are defined as those which are applied to a base item where a significant extra burden of work similar to the nature of work in the base item is placed upon the Contractor e.g. extra over excavation for excavation in hard material. The quantities billed must be for work which is included in quantities for the base item; this is often a source of error. For example if there are 1000 m³ of hard material then that 1000 m³ should be included in the appropriate excavation item. Note also that the Item Coverage for the extra over item consists of not only its own Item Coverage but also that of the appropriate base item.

*Series 100—Preliminaries* Details of services (and supplies) and services (and supplies) diversions should be listed in Appendix 1/16 which should be included in the Specification, see the example given near the end of Series 100 in Volume 2: 'Notes for Guidance on the Specification for Highway Works'.

Service locations are notoriously inaccurate and often cause substantial delay and disruption. Alterations to services by undertakers or their Contractors can also be a source of delay and disruption.

*Series 500—Drainage and Service Ducts* Tabulated billing is very convenient but the actual average depths are often substantially different from the table.

*Series 600—Earthworks* The earthworks balance is often rather different from that which is suggested in the bill of quantities and it is this item which is probably the source of most claims. Often, it is very difficult for the Engineer to get the proportions of different types of acceptable/unacceptable material correct and equally, it is often the case that this change in balance can upset the Contractor's programme or incur extra costs.

The Engineer is required to include a Roadworks Earthworks Schedule and a Structures Earthworks Schedule, where appropriate, as part of the contract documentation. These should be maintained and checked by the Contractor as work proceeds to ensure that his tender assumptions are not jeopardised.

### 3.6.4. Section 3—Library of Standard Item Descriptions for Highway Works

Each of the sub-headings in Section 1, Method of Measurement for Highway Works contains an Itemisation Table from which it is possible to produce the Item Descriptions. This section lists the options available for Item Descriptions. It makes the proper compilation of bills of quantities very simple. As before, it is set out using the standard 26 Series but in this section not only is Series 2700 added but Series 2800: Provisional Sums and Prime Cost Items is appended. Having noted that, Series 2700 is 'Not taken up' as are Series 800, 900, 1000 and 2600.

The use of an example is the best way of explaining how the system operates. On page 1 of Series 100: Preliminaries, item number 1 reads as follows.

1* of principal offices for the Engineer 2*           item

Page 3 of the same sub-section gives 1* as

|   |   |   |
|---|---|---|
|   | (i) | =Erection |
|   | (ii) | =Servicing |
|   | (iii) | =Dismantling |
| and 2* as | (o) | =no entry |
|   | (i) | =provided by the Employer |

It can be seen how Item Descriptions can be compiled with ease. The system continues for all the other Series.

This section also contains a very important note under paragraph 2 Amendments to the Library. It concerns what are called 'rogue items'. These are for types of work which fall outwith those items which are mentioned in the Method of Measurement. It states that rogue items 'should be drafted on the same principles as the Library and inserted as necessary in the Bill of Quantities'. It goes on to state that as 'in the case of the MMHW' new items which are found to be required often should be forwarded to the Engineering Policy and Programme Division, for possible inclusion. Nowhere in either Section 1, the Method of Measurement for Highway Works or Section 2, Notes for Guidance on the Method of Measurement for Highway Works does it say that rogue items should be compiled on the same principles. Nevertheless, it is difficult to see how they could be included unless the same approach is employed. Rogue items are often ill-conceived, badly worded and poorly implemented, and Contractors should check these carefully to see whether or not they have been properly produced.

## 3.7. Claims associated with the *Manual of Contract Documents for Highway Works*

### 3.7.1. General

The *Manual of Contract Documents for Highway Works* is a superb system. When the rules are properly applied, the Contractor can make due allowance in his rates for those elements that combine to comprise the Works. Where the system has not been applied properly then the Contractor may well find that his rates are inadequate to cover the costs of the work comprised in one or more items. In many cases this will simply result in a re-rating exercise. In more serious cases, say where a number of items are affected or where it involves an item which has a large quantity attached, then the effects may be much more significant. This is particularly true where the nature of the work is in dispute such that it takes longer to execute and affects the programme causing delay or disruption. The *Manual of Contract Documents for Highway Works* is the most common means employed for preparing highway contracts. When a document is prepared on behalf of the Highways Agency, it is then sent to the Agency for their approval. Any checks on contracts prepared for non-trunk road improvements are subject only to internal checks (if any) within the organisation preparing the contract. Experience suggests that this situation is much more likely to produce a document which is not in strict accordance with the requirements of the *Manual of Contract Documents for Highway Works*. Furthermore, the likelihood of non-compliance appears to increase when non-standard items are involved.

When two parties contract to carry out Works of construction, the Engineer assumes that the Contractor has studied all the elements of the contract and is fully aware of all the details pertaining to the contract. This is simply not the case. The Contractor normally has very little time to become familiar enough with the contract in order to price the tender. Notwithstanding, once the contract is signed, the express conditions, by and large, will apply and the courts would normally take the view that the Contractor was aware of his obligations before the contract was signed and accepted the conditions which form part of that contract. It is almost always the case that the Contractor only has the luxury of time available after the contract has been signed. Engineers will often say to the Contractor that he must have been aware of a perceived shortcoming in the contract documentation and question why the matter was not brought up for clarification before the contract was signed. There are four reasons why he may not have done so.

(a) The contractor may well not have realised any potential conflict at the time of tender because of the usual rush to obtain prices and, nowadays, to meet the myriad of other pre-tender requirements.
(b) The contractor did not wish to be seen to be difficult, particularly at the pre-acceptance stage.

(c) The contractor may have seen it as a claim opportunity during execution.
(d) There was no need for the Contractor to do so—there was no contractual relationship nor was there any other requirement to raise any areas of concern.

The function of this section is to set out some ground rules for examining whether or not the contract has been compiled properly and, if not, to objectively assess the effects thereof.

### 3.7.2. Claims emanating from the Specification

The function of the Specification is to dictate how the Works are to be executed in terms of both the methods to be adopted and/or the end result to be achieved. It is always useful to check that the Specification has been properly presented in the contract and a few simple checks should confirm that this is the case.

At the start of a contract, particularly a large one, it is worth going through Section 000 of Volume 1: 'Specification for Highway Works' and checking

(a) the contents of Appendix 0/5 if the contract is supervised by the Scottish Office Industry Department, the Welsh Office or the Department of the Environment for Northern Ireland and hence which Clauses apply to the contract
(b) the contents of Appendix 0/3 and hence the Numbered Appendices which apply to the contract.

Table 3.1 lists the requirements of the generic specification documents and is useful as a check list.

Except for Series 800, 900 and 1000, each of the remaining Series has one or more Appendices and samples of each can be found in Volume 2: 'Notes for Guidance on the Specification for Highway Works'. They should be supplied in accordance with the above as required, duly amended. If they are not provided or are provided incorrectly completed then the Contractor must consider what the ramifications of these shortcomings are and claim or not claim accordingly.

### 3.7.3. Claims emanating from the Bill of Quantities

Section 3 of Volume 4: 'Bills of Quantities for Highway Works' consists of the Library of Standard Item Descriptions for Highway Works. The Contractor should check that these comply with the

Table 3.1. (below, facing and overleaf). List of Key Appendices

| Requirement | Description | Where stated | Notes |
|---|---|---|---|
| Appendix A | Quality management schemes | NG 003.1, 003.2 | |
| Appendix B | Product certification schemes | NG 003.1, 003.2 | |
| Appendix C | British Board of Agrément Roads and Bridges Certificates | NG 003.1, 003.2 | |
| Appendix D | Statutory type approval | NG 003.1, 003.2 | |
| Appendix E | Departmental type approval/registration | NG 003.1, 003.2 | |
| Appendix F | Publications referred to in the specification | NG 003.3 | |
| Appendix G | Petrographical examination of aggregates for alkali–silica reaction | NG 003.1, 003.2 | |
| Appendix 0/1 Additional | List of contract-specific Clauses, Tables and Figures suffixed by AR | NG 003.4, 003.5 003.10 | |
| Appendix 0/1 Substitute | List of contract-specific Clauses, Tables and Figures suffixed by SR | NG 003.4, 003.5 003.10 | |
| Appendix 0/1 Cancelled | List of contract-specific Clauses, Tables and Figures suffixed by CR | NG 003.4, 003.5 003.10 | |
| Appendix 0/1 Additional | Actual contract-specific Clauses, Tables and Figures | NG 003.4, 003.5 003.10 | |

## 7TH EDITION OF THE MANUAL OF CONTRACT DOCUMENTS | 179

*Table 3.1—continued*

| Requirement | Description | Where stated | Notes |
|---|---|---|---|
| Appendix 0/1 Substitute | Actual contract-specific Clauses, Tables and Figures | NG 003.4, 003.5 003.10 | |
| Appendix 0/2 | Contract-specific minor alterations to existing Clauses, Tables and Figures | NG 003.4, 003.5 003.11 | Using 'delete and insert' or 'add at end of Clause' |
| Appendix 0/3 List A | List of Numbered Appendices referred to in the specification and included in the contract taken from Volume 1: 'Specification for Highway Works' | NG 003.4, 003.5 003.12 | Where a Numbered Appendix is not used it should be marked 'Not Used'. Check carefully the symbols E, E/C, E/T, C, I and P and the implications thereof |
| Appendix 0/3 List B | List of Numbered Appendices referred to in the Specification and included in the contract specially devised for the contract | NG 003.4, 003.5 003.12 | Numbering should commence at 1/70 to avoid confusion with future additional national Numbered Appendices |
| Appendix 0/4 Table 1 | List of contract-specific drawings supplied to each tenderer | NG 003.4, 003.5 003.13 | |
| Appendix 0/4 Table 2(i) | List of standard drawings supplied to each tenderer | NG 003.4, 003.5 003.13 | |
| Appendix 0/4 Table 2(ii) | List of standard drawings inspected by tenderers | NG 003.4, 003.5 003.13 | |
| Appendix 0/4 Table 2(iii) | List of drawings brought into the contract of reference | NG 003.4, 003.5 003.13 | |

Table 3.1.—continued

| Requirement | Description | Where stated | Notes |
|---|---|---|---|
| Appendix 0/5 Additional | List of special national alterations to Clauses, Tables and Figures suffixed by A | NG 003.4, 003.5 003.14 | Clauses should commence with the 50th number in the series or be numbered sequentially with the national alteration Clauses in the SHW |
| Appendix 0/5 Substitute | List of special national alterations to Clauses, Tables and Figures suffixed by S | NG 003.4, 003.5 003.14 | |
| Appendix 0/5 Cancelled | List of special national alterations clauses, Tables and Figures suffixed by C | NG 003.4, 003.5 003.14 | |
| Appendix 0/5 Additional | Actual special national alterations to Clauses, Tables and Figures | NG 003.4, 003.5 003.14 | |
| Appendix 0/5 Substitute | Actual special national alterations to Clauses, Tables and Figures | NG 003.4, 003.5 003.14 | |

wording set out in this Section. Where it does not then the Contractor has to consider what the ramifications of these shortcomings are and claim or not claim accordingly. The effects of variation in the stated and actual quantities are more properly a matter to be considered under the terms of the *5th Edition of the ICE Conditions of Contract* and are covered in detail in Chapter 2.

In the case of non-compliance of non-standard items, this should be considered in the light of the advice contained in the next part of this chapter.

### 3.7.4. Claims emanating from the Method of Measurement

It is worthwhile spending some time looking at individual items to see whether they comply with the rules established in the *Manual of Contract Documents for Highway Works*. According to the general principles set out in Section 1, Chapter II of Volume 4: 'Bills of Quantities for Highway Works' the combination of the Sub-Heading, the Item Description and the Item Coverage for the appropriate Marginal Headings in conjunction with the matters set out under General Directions in the Preambles to Bill of Quantities in Section 1, Chapter III of Volume 4 must identify the work covered in respect of any particular item. A table such as Table 3.2 is useful in identifying non-conformances and their effects.

Table 3.2 is used as follows. All the item details are entered into the first and second columns. The third column indicates whether the item is a standard item from the Method of Measurement or whether it is a non-standard item. The next column indicates whether there is the normal information required under the Marginal Heading i.e. Unit, Itemisation Table and Item Coverage. The next column indicates whether the extent of the Item Coverage is adequate. Next is the compliance/adequacy of the Item Description. The final column should be used to detail any non-conformances and their effects.

In the example above, the first three items, standard items from the Method of Measurement, comply in all respects. The next three, non-standard items, have various problems. Item 2·246 has an Itemisation Table, an Item Coverage but the Item Description does not comply with the Itemisation Table. Nevertheless, the item meaning is still fairly clear. Item 7·003 has a deficient Item Coverage in that several elements of work are missing namely trimming the pothole, spreading the material and compaction, all elements which appear, incidentally, in the standard surfacing Item Coverages, a lesson there for someone (actually they failed to import Series 700, para-

Table 3.2. *Examination of Bill of Quantities items*

| Item no. | Brief description | S/NS | Compliance — Marginal heading | Compliance — Item coverage | Compliance — Item description | Non-conformance/Effect/Notes |
|---|---|---|---|---|---|---|
| 1.001 | Erection of offices for the Engineer | S | ✓ | ✓ | ✓ | Nil |
| 1.002 | Servicing of offices for the Engineer | S | ✓ | ✓ | ✓ | Nil |
| 1.003 | Dismantling of offices for the Engineer | S | ✓ | ✓ | ✓ | Nil |
| — | — | — | — | — | — | — |
| 2.246 | Remove litter/debris from verges etc. | NS | ✓ | ✓ | ✗ | Meaning still fairly clear/No effect |
| — | — | — | — | — | — | — |
| 7.003 | Temporary repair to pothole | NS | ✓ | ✗ | ✓ | Trimming, spreading, compaction not included/Re-rate |
| — | — | — | — | — | — | — |
| 29.050 | Provide and maintain salt storage depot(s) etc. | NS | ✗ | ✗ | ✗ | No itemisation table, no item coverage, meaning wholly unclear/Re-rate after preamble received |

S—Standard item from Method of Measurement
NS—Non-standard item

graph 9, to the Item Coverage). The third example has no Marginal Heading Units, Itemisation Table or Item Coverage, a recipe for a claim and rightly so.

The fact that there is no Marginal Heading i.e. no Units, Itemisation Table or Item Coverage does not, of itself, mean that there is a valid claim. Assessment then comes down to what the objective observer would take the item to mean. If the Item Description is such that the work involved in the item is self-evident then the Contractor is unlikely to have a valid claim. Where there are a number of reasonable interpretations then any of those adopted by the Contractor is deemed to be valid and a claim is likely to be successful. Nevertheless, the Engineer would be well advised to make sure that such a situation does not occur and, as has been seen, this situation is relatively easy to avoid if the rules of compilation are followed.

The claim in Chapter 5 is based around the occurrence of what the Contractor suggests is an adverse physical condition. The Contractor draws on various parts of the *Manual of Contract Documents for Highway Works* to support his case; an approach very common in claims on highway contracts.

## 3.8. References

1. **Money B. and Hodgson G.** *Manual of Contract Documents for Highway Works—A User's Guide & Commentary.* Thomas Telford, London, 1992, **1** and **2**.
2. **Money B. and Hodgson G.** *Manual of Contract Documents for Highway Works—1993/1994 Amendments—A User's Guide & Commentary.* Thomas Telford, London, 1996.
3. **The Federation Of Civil Engineering Contractors.** *Schedule Of Dayworks Carried Out Incidental To Contract Work.* The Federation Of Civil Engineering Contractors, London, 1990.

# 4

## The claims process

### 4.1. Preamble

Other than those which are dictated by the *ICE Conditions of Contract* in terms of notification requirements etc. there are no set rules as to how the claims resolution process should be conducted unless the claims go to a formal arbitration where there are very strict procedures to be followed. However, any steps which lead to a mutually acceptable resolution and avoid arbitration are to be encouraged.

This chapter looks at how the process develops and suggests why resolution at an early stage is to the benefit of all involved. If the parties cannot reach agreement then the process escalates through to arbitration and perhaps even court action before the claim reaches its conclusion. The proper presentation of claims is vital if the Contractor is to maximise the possibility of achieving a satisfactory conclusion.

### 4.2. Foreplay

A claim may arise due to circumstances which, the Contractor believes, are clearly covered by the *Conditions of Contract*. In other cases, the Contractor will be unsure of his grounds and will seek clarification from the Engineer. Often this is an attempt either to gauge the Engineer's resolve or to ascertain how he regards a particular issue. Such requests may well take the form of a letter asking why a particular interpretation is invalid; it may point out that there are two (or more) ways in which a contractual obligation can be met; or it may ask how a particular element of the Works or obligation within the contract is to be paid. These letters should ask the Engineer to substantiate his reasoning by reference to specific provisions of the contract documentation. This is often the first step in the claims process. If the response from the Engineer convinces the Contractor that his tentative probing of the contract documents is repudiated with justification then the Contractor would be wise to take the matter no further. However, there is an automatic inclination in many Engineers to rebuff a claim on the (unstated) basis that he prepared the contract documents and, therefore any blight thereof is inherent criticism of the Engineer himself. The Contractor should bear this in mind and be as circumspect as possible. As will be seen, the longer the process, the greater the tendency for the parties to adopt entrenched positions and, in such circumstances, there is less likelihood of compromise. It is worth considering the Engineer's response in depth and taking some time to do so. What initially seems to be a justified repudiation of a claim can, two days later, seem substantially flawed.

This process can continue either through correspondence or via discussion at meetings. Projects are usually controlled via regular progress meetings. It is often useful to schedule separate commercial meetings at which all questions about financial matters can be considered and discussed. This frees the progress meetings of discussions related to payment matters which often occupy a substantial amount of time. During these discussions, there are several fundamental rules which should be observed. These meetings should be conducted in a civilised manner with all present according dignity and courtesy to all others attending the meeting. Interrupting the speaker really is inexcusable and should be discouraged by the senior person present (preferably before the meeting). Discussions on financial matters will often produce passionate argument on both sides but cool logic is liable to be more effective. The meetings should be the subject of an agenda which is likely to be provided by the Contractor. Copies of all documents which are contractual, even arguably so, should be available for reference purposes. It is helpful if those relying on particular elements of the contract documents can photocopy the relevant parts for production at the meeting. Finally all should prepare themselves in advance of the meeting as far as possible. Care should be taken before making unprepared statements—there is nothing more embarrassing than having your point destroyed in front of several colleagues. Verbatim minutes are impractical at such meetings but notes of salient points are often useful for future reference. The participants should try to enjoy themselves; the intellectual challenge can be fun. Finally all should remember that there are many matters in life which are infinitely more important and it is really not worth getting upset about an issue—besides being calm is more likely to produce success than a show of temper.

Compromise by both sides is a wise policy. It does not matter to the Contractor whether the claim that was judged to be the most sound was utterly rejected by the Engineer. What is important is that the Contractor receives a fair payment for his efforts; the means of doing so is, to a certain extent, academic. A Contractor who has executed the engineering side of the contract with skill and expedition will often find that claims are more favourably received by the Engineer. Compromise is very often necessary for the successful resolution of disputes. A Contractor who does not expect to compromise is not acquainted with reality. Both parties will want to walk away feeling that they have not capitulated on all points and that an equitable solution on the part of all parties was reached. Equally the Engineer should bear in mind the fact that the earlier a claim is settled, the cheaper it tends to be. While it is understood that this may encourage a Contractor who is claims orientated beyond that which is warranted by the contractual circumstances, it is very often the time at which the Contractor has not fully developed arguments and is least sure of the case. Another fact in favour of early settlement is that the parties have not adopted an entrenched position. As time passes and discussions or exchanges of letters take place it becomes ever more difficult to back down. All parties should remember that no matter what the strength of one's convictions about the merits of their view of the contract, the outcome of an arbitration is never certain. Arbitrations are almost always expensive affairs and should be avoided if possible. Sometimes, however, a Contractor's claims are ill-founded or the Employer will not sanction any compromise on what are fully justified claims and, as a consequence, there is no alternative to arbitration.

This Chapter and the next concentrate on the proper presentation of claims. By the time a contract reaches this stage, the arguments have normally been fully aired at meetings and in correspondence. So, the purpose of presenting the claims in a formal manner is to give the Engineer the opportunity to see the entire claim from contractual inception to a statement of its financial consequences. The Engineer will also attempt to gauge the strength of the Contractor's resolution and determination in pursuing the claim because, as has been seen, the outcome of the process is often determined not by the strength of the case of the parties but by the parties perception of the strength of their case. Contractors will, on occasions, pursue a weak case because they feel that the Engineer will settle at some stage; this view is not always borne out by experience.

Notwithstanding, a well presented case which is easy to follow and structured logically can only enhance the possibility of settlement. The next Chapter sets out a claim in a manner which is based on just such an approach.

## 4.3. Presentation of claims

The claims document/s should be packaged professionally. While it may not be a wise approach, it is a fact that many people do judge a book by its cover. All parts of the document/s should be of the highest quality: the binding, the printing, the syntax and the argument. Nowadays, the preparation of high quality reports is within the reach of the most modest office. Computers with good quality software and sophisticated graphics capabilities are no longer expensive items of equipment. When connected to a laser printer capable of printing text or graphics at high resolution they can produce documents which only a few years ago seemed to be the sole province of a professional printing bureau. The front and back covers can be laminated and the entire document edge perforated to produce a finished article which sets the scene for what should be an equally impressive content. Chapter 5 presents a claim in its final format.

# 5

## Model claim

This Chapter consists of a claim made by a Contractor engaged on a roadworks contract. Notes on various aspects of the claim are printed in italics.

### A1069—BIGSVILLE TO LITTLESVILLE CONSTRUCTION OF DUAL CARRIAGEWAY

### CLAIM FOR EXTENSION OF TIME AND COSTS

### JANUARY 1997

UNLIMITED CONTRACTING LTD
28 THE YARD
NEWTOWN INDUSTRIAL ESTATE
NEWTOWN NT9 9ZZ

EMPLOYER:

MEDIUM COUNTY COUNCIL
COUNTY HALL
NEWTOWN NT1 1AA

ENGINEER:

MR R SHARPE
SHARPE & SHARPE
CONSULTING ENGINEERS
2 THE MEWS
NEWTOWN NT2 1BB

# Contents

1. **Preamble**

2. **Contract details**

3. **Nature of claim: claim related to adverse physical condition in the form of unacceptable material below the foundation of a box culvert**

4. **Evaluation of claim**

**Appendix 1.** Sketch of the layout of the Works

**Appendix 2.** Original Clause 14(1) programme

**Appendix 3.** List of drawings

**Appendix 4.** Correspondence related to claim

**Appendix 5.** Programme showing actual progress

**Appendix 6.** Dayworks sheets

**Appendix 7.** Non-productive overtime costs

## 1. Preamble

This contract consists of the upgrading of 2 kilometres of single carriageway road to a dual carriageway including the construction of a roundabout and a box culvert. The purpose of the latter is for use as a cattle culvert being accommodation works for a farmer who owns land on both the north and south sides of the carriageway. Principal quantities include the excavation of some 500 m$^3$ of topsoil and 10 000 m$^3$ of acceptable material including some 1 500 m$^3$ of hard material, the importation of some 5 000 m$^3$ of hard material, the construction of some 6 500 m of filter drains and some 6 000 m of surface water drains and the laying of some 30 000 m$^2$ of roadbase, 32 000 m$^2$ of basecourse and 38 000 m$^2$ of wearing course. Appendix 1 depicts a diagram showing a sketch layout of the Works.

Invitations to tender were issued to a number of contractors including the claimant. The priced tender documents including a Form of Tender dated 6 March 1996 were returned to the Employer on 7 March 1996, the last day for submission of tenders. A Form of Agreement was signed on 24 April. Under cover of his letter dated 6 May the Engineer notified the Contractor that the Date for Commencement of the Works was to be 20 May. Since the Time for Completion was 26 weeks, this fixed the date for completion of the whole of the Works as 17 November 1996.

We submitted our programme and a general description of the arrangements and methods of construction in accordance with the requirements of Clause 14(1) of the Conditions of Contract on 14 May and it is reproduced in Appendix 2. The Engineer sought verbal clarification of a number of points in relation thereto and thereafter under cover of his letter dated 21 May approved the programme pursuant to Clause 14(1) of the Conditions of Contract.

*Note. The Contractor has cited his programme in this Claim and is thus suggesting that it has the status of* prima facie *evidence that encountering this physical condition resulted in delay. As discussed earlier, the standard contract (as per the 5th Edition of the ICE Conditions of Contract) does not require the production of a programme along with the tender. Furthermore, the approval thereof by the Engineer does not necessarily give it that status. If other events demonstrate that the timescales adopted therein were reasonable and there is an absence of any other information to the contrary then the Engineer may have difficulty in repudiating its relevance.*

Work commenced on site on 20 May. A Certificate of Completion for the whole of the Works was issued on 12 November by the Engineer pursuant to Clause 48(1) and giving the Date of Completion as 5 November 1996. No Final Account has yet been issued. As at the date of preparation of this Claim, the Engineer has certified payment of £617 283·69, all of which has been paid by the Employer.

*Note. No final account is required until three months after the issue of the Maintenance Certificate (probably due 5 November 1997).*

## 2. Contract details

The Employer is Medium County Council, County Hall, Newtown NT1 1AA.

The Engineer is Mr R Sharpe BSc, CEng, FICE, FIHT, Sharpe & Sharpe, Consulting Engineers, 2 The Mews, Newtown NT2 1BB.

The Contractor is Unlimited Contracting Ltd, 28 The Yard, Newtown Industrial Estate, Newtown NT9 9ZZ.

The total payable based on the tender Bill of Quantities is £632 522·17.

The contract documents consist of:-

> the document containing the Forms of Tender, Agreement by Deed and Bond, Conditions of Contract (as priced), Specification and Bill of Quantities

which imports into the contract

(a) the Drawings listed in Appendix 3
(b) the 5th Edition of the ICE Conditions of Contract (June 1973) (Revised January 1979) (Reprinted January 1986) as extended and amended within the contract
(c) the 7th Edition of the Specification for Highway Works dated December 1991 as amended August 1993 and August 1994 as extended and amended within the contract
(d) Section 1 of Volume 4: Bills of Quantities for Highway Works, Method of Measurement for Highway Works dated December 1991 as amended August 1993 and August 1994 as extended and amended within the contract.

## 3. Nature of claim: claim related to adverse physical conditions in the form of unacceptable material below the foundation of a box culvert

We encountered unacceptable material as defined in Volume 1: Specification for Highway Works while excavating the foundations to the box culvert which formed part of the Works. We consider this event to constitute an adverse physical condition which could not reasonably have been foreseen by an experienced contractor. Accordingly, under the terms of Clause 12(1) of the 5th Edition of the ICE Conditions of Contract, we gave the Engineer notice that we had encountered such condition and that additional costs would be incurred. Further correspondence followed and is summarised in Table 1. All letters are reproduced in their entirety in Appendix 4.

Note. It is worth reading the paragraph contained in Section 2 of Volume 4: 'Bills of Quantities for Highway Works', Notes for Guidance on the Method of Measurement for Highway Works. Paragraph 7 in Series 600: Earthworks is headed Hard Material and states that if 'the material found during the course of construction is that which was shown at the time of tender, or could be ascertained by inspection in accordance with Clause 11, then admeasurement should follow the same designations irrespective of the actual hardness of the material. If the material found in the course of construction is not as described in the tender documents or apparently by inspection, the Contractor may raise a claim under Clause 12 of the Conditions of Contract. It will then be for the Contractor to demonstrate that the material could not reasonably have been foreseen and that extra costs had arisen, according to the terms of that Clause'. Notwithstanding the fact that this relates to hard material, the underlying principle applies. If

(a) the situation encountered during the course of construction is not in conformity with the contract or apparent in accordance with Clause 11, and
(b) the Contractor demonstrates that it could not reasonably have been foreseen, and
(c) he is delayed and/or incurs extra cost then the Contractor is entitled to payment and/or extension of time as per Clause 12.

Our letter dated 17 September confirms that we continue to hold the opinion that this event constitutes an adverse physical condition pursuant to Clause 12 of the Conditions of Contract. Accordingly, we seek our rights in relation thereto, viz

(a) the granting of an extension of time to take account of the delay suffered by us in executing the Works and reasonable costs in relation thereto
(b) payment of such sum as represents the reasonable cost of carrying out the additional work done and additional Constructional

Table 1. Log of correspondence to claim

| Letter no. | Date of letter | From/To | Contents |
|---|---|---|---|
| C1 | 12 July 1996 | Contractor/Engineer | Notice of encountering adverse physical condition and proposals for dealing with same |
| C2 | 15 July 1996 | Engineer/Contractor | Request for provision of estimate, without prejudice |
| C3 | 17 July 1996 | Contractor/Engineer | Provision of requested estimate |
| C4 | 17 July 1996 | Eng. Rep./Contractor | Confirmation of verbal instruction to extend depth of excavation |
| C5 | 7 August 1996 | Contractor/Engineer | Notice of claim |
| C6 | 8 August 1996 | Engineer/Contractor | Notice that the Engineer's decision is that the conditions were reasonably foreseeable by an experienced contractor |
| C7 | 16 August 1996 | Contractor/Engineer | Attempt to persuade the Engineer that the physical condition was not reasonably foreseeable by an experienced contractor |
| C8 | 27 August 1996 | Engineer/Contractor | Further support for Engineer's decision |
| C9 | 17 September 1996 | Contractor/Engineer | Repeated notice of claim by Contractor |

Plant used which would not have been done or used had such condition not been encountered together with a reasonable percentage addition thereto in respect of profit

(c) payment of damages associated with the consequences of the failure of the Engineer to certify an extension of time and costs as cited above within a reasonable period of the occurrence of the condition.

## 4. Evaluation of claim

### 4.1. Costs associated with extension of time

Appendix 5 shows a revised version of the original programme which shows the actual periods for construction. As can be seen, the original time periods which were envisaged at the time of presentation based on the contract information were met. The only major deviation is that which relates to the excavation below the foundation to the box culvert. An extra element of work has been inserted to represent the additional work.

Originally the formwork to the base of the box culvert was programmed to commence on 15 July. In fact it did not commence until 6 August. The extra work took 17 days which was executed on the basis of a 5 day working week which translates into 22 days in terms of the contract.

This meant that the site establishment and supervisorial staff had to be maintained on site for 22 days longer than would have been the case if the soft material below the foundation had not been encountered. These costs are shown below.

Furthermore, this period would attract payment of head office overheads and profit.

*4.1.1.  Staff*  Table 2 shows the costs associated with staff payments for the extended period.

*4.1.2.  Site accommodation etc.*  Table 3 shows the costs associated with the provision of site accommodation for the extended period.

*4.1.3.  Head office overheads and profit*  Our accountants have audited our accounts for the last three years and have certified that our head office overheads and profits as a percentage of turnover are

*Table 2. Prolongation costs for managing staff*

| Occupation | Rate/Week | Period | Cost |
|---|---|---|---|
| Project Director—part time only | £800·00 | (say) 11 days | £1257·14 |
| Site Agent | £575·00 | 22 days | £1807·14 |
| Setting Out Engineer | £350·00 | 22 days | £1100·00 |
| Measurement Engineer | £425·00 | 22 days | £1335·71 |
| General Foreman | £350·00 | 22 days | £1100·00 |
| Technical Clerk | £300·00 | 22 days | £942·86 |
| Cleaner | £150·00 | 22 days | £471·43 |
| Watchman | £150·00 | 22 days | £471·43 |
|  |  | Total | *£8485·71* |

Note. These rates include all employee overheads e.g. NI, sick pay, holiday pay, car provision etc.

*Table 3. Prolongation costs of site accommodation*

| Item | Rate/Week | Period | Cost |
|---|---|---|---|
| Hire of Contractor's site offices | £850·00 | 22 days | £2671·43 |
| Hire of associated equipment (computers, surveying equipment, photocopier, furniture) | £350·00 | 22 days | £1100·00 |
| Hire of Engineer's site offices | £450·00 | 22 days | £1414·29 |
| Hire of associated equipment (computers, surveying equipment, furniture) | £250·00 | 22 days | £785·71 |
| Electricity, gas, telephones | say £300 | – | £300·00 |
| Security | £100·00 | 22 days | £314·29 |
| Small tools | £50·00 | 22 days | £157·14 |
| Additional insurance | £1500·00 | – | £1500·00 |
|  |  | Total | *£8242·86* |

| | |
|---|---|
| 1993–1994 | 11·26% |
| 1994–1995 | 9·81% |
| 1995–1996 | 10·67% |

The average of the above is 10·58%.

*Note.* All probative information should be provided as part of the claim, for example a statement from the Contractor's accountants to verify the information given above. This saves wasting time (which, of course, costs money) when the Engineer asks for proof of same which would, in effect, delay payment.

Over 22 days the contribution would have been

$$\frac{10{\cdot}58\% \times \text{Contract price} \times 22 \text{ Days}}{\text{Contract period}}$$

$$= \frac{10{\cdot}58\% \times 617\,283{\cdot}69 \times 22}{26 \times 7}$$

$$= £7894{\cdot}45$$

*Note.* This approach is often called the Hudson or Emden Formula. Where one of the figures for the previous three years is abnormally high or low then it may be set aside on the basis that it is untypical for the performance of the Contractor.

Total amount of this element of the claim

$$= 8485{\cdot}71 + 8242{\cdot}86 + 7894{\cdot}45$$

$$= £24\,623{\cdot}02$$

### 4.2. Costs of carrying out the work

The cost of carrying out the excavation and backfilling has been paid for at billed rates. We contend that this work ought to have been the subject of an ordered variation pursuant to the terms of Clause 51(1) of the Conditions of Contract. Accordingly, it should have been evaluated under the terms of Clause 52(3) of the Conditions of Contract.

*Table 4. Dayworks for excavation and replacement of unacceptable material*

| Daywork no. | Period covered | Value |
|---|---|---|
| 51153 | 15 July 1996–20 July 1996 | £4328·95 |
| 51156 | 21 July 1996–27 July 1996 | £11 320·11 |
| 51157 | 28 July 1996–3 August 1996 | £8381·13 |
| 51162 | 4 August 1996–6 August 1996 | £2579·25 |
| | Total = | £26 609·44 |

Appendix 6 contains copies of Dayworks sheets which evaluate the work. The total in each is given in Table 4.

A total of £8683·45 has been certified by the Engineer against this work. This sum was calculated at billed rates which, as explained earlier, we do not believe to be appropriate in the circumstances. The Employer has paid this sum and so the net amount due under this heading is £26 609·44 − £8683·45 i.e. £17 925·99.

### 4.3. Damages due to the failure of the Engineer to certify extension of time

These come under three headings.

(a) Costs associated with working non-productive overtime to achieve contractual Date for Completion.
(b) Finance charges payable to fund the extra work.
(c) Costs of preparing claim.

*4.3.1. Costs for non-productive overtime* When the existence of the unacceptable material below the foundation to the box culvert became apparent, we believed that a Variation would be ordered and there would follow an appropriate extension of time and payment of reasonable costs. We were very disappointed that the Engineer felt unable to support this approach. Given the fact that we were unable to persuade the Engineer to change his mind we had to consider what action to take in the circumstances. This event had added, in our view, some 22 days to the contract and we had a contractual obligation to complete by the Date for Completion. We take pride in meeting our contractual obligations. Furthermore, we are a commercial organisation which has to consider the interests of our shareholders and, aware that liquidated damages had been set at £260 per day, we had to ensure that we did not incur these charges. Accordingly we had to take measures to accelerate the execution of the Works in order to ensure completion within the permitted period.

There were a number of options available to us in order to achieve early completion. We prepared our tender on the basis of working a 5-day week with each day having nine working hours. Up to 12 July when we first encountered the problem under the foundation we had, within reasonable limits, worked to our programme submitted in accordance with Clause 14(1) of the Conditions of Contract and although we were confident that we would complete the Works within the contract period, we could not recover 22 days without operating under a different regime. We did consider increasing the manpower which was utilised on the site but decided that this was not the entire solution and therefore decided that we had no option but to work overtime also in order to ensure contracted completion. Table 5 lists the projected and actual periods taken for the operations on site.

*Table 5. Comparison of actual and programmed execution times*

| Operation | Actual time in days | Programmed time in days |
|---|---|---|
| *Roadworks* | | |
| Site set up | 1 | 1 |
| Strip topsoil | 5 | 5 |
| Bulk excavation | 40 | 40 |
| Surface drainage | 18 | 20 |
| Filter drainage | 22 | 25 |
| Sub-base | 30 | 30 |
| Roadbase | 15 | 35 |
| Basecourse | 10 | 15 |
| Wearing course | 8 | 15 |
| Road mkgs/Tr. signs | 6 | 5 |
| Clean up site | 1 | 1 |
| Total | 156 | 192 |
| *Box culvert* | | |
| Excavate foundations | 4 | 3 |
| Excavate below foundations | 18 | N/A |
| *Base* | | |
| Formwork | 2 | 5 |
| Fix steel | 4 | 5 |
| Pour concrete | 5 | 5 |
| Strike formwork | 1 | 2 |
| *Walls* | | |
| Formwork | 5 | 5 |
| Fix steel | 4 | 5 |
| Pour concrete | 5 | 5 |
| Strike formwork | 1 | 2 |
| *Soffit* | | |
| Formwork | 5 | 5 |
| Fix steel | 4 | 5 |
| Pour concrete | 5 | 3 |
| Strike formwork | 1 | 2 |
| Backfill | 2 | 4 |
| Tidy structure | 1 | 1 |
| Total | 49 (+18) | 57 |

We did increase working hours for the structural gangs and reduced the projected time period to 49 days; a reduction of 8 days. We also increased the working hours of the roadworks gangs from the start of the roadbase operations (there was no point in doing so with sub-base since its completion was not possible until the structure had been backfilled). It was not possible to commence roadbase earlier since it was laid by a sub-contractor and he was committed to Works on other contracts.

The extra costs which we faced as a result of the actions described above came about from extra charges levied by the sub-contractor who was instructed by us to work extra hours and our own extra costs as a result of paying overtime rates to our own staff. The costs associated with these non-productive hours are shown in Appendix 7 and total £6551·53 + £5456·44 = £12 007·97.

*4.3.2. Finance charges* In order to fund the extra work we have had to borrow money to cover the costs. Table 6 sets out the nature of the charge and the dates from which it should have been paid and therefore the date from when interest is due.

Note. The date chosen from which finance charges should run are as follows

(a) extension of time—date of Engineer's letter repudiating variation
(b) costs of execution—date when sums would have been paid in the monthly statement and
(c) damages—halfway between the dates when the Contractor's own staff were paid and when the surfacing sub-contractor was paid.

The finance charges accrue on the basis of 2 per cent above our bank's base rate as charged in line with the procedure set out in Clause 60(6) of the 5th Edition of the Conditions of Contract.

Amount of Finance Charges up to 31 January 1997
= £3159·15

Table 6. *Dates when finance charges became due for elements of claim*

| Head of claim | Amount | Date due |
| --- | --- | --- |
| Extension of time | £24 623·03 | 8 August 1996 |
| Costs of executing the work | £17 925·99 | 25 September 1996 |
| Damages due to non-certification | £12 007·97 | 14 September 1996 |

*4.3.3.   Cost of preparing claim*   We have paid a consultant the sum of £6462·50 to prepare this claim.

   *Note. It is suggested that it is always wise to insert a note in the file stating that engaging a consultant was with a view to the claim going to arbitration.*

   Total amount of this element of the claim

$$£12\,007·97 + £3159·15 + £6462·50 = £21\,629·62$$

*4.4.   Summary of claim*

(a) Costs associated with extension of time = £24 623·02
(b) Costs of carrying out the work = £17 925·99
(c) Damages due to non-certification = £21 629·62

Total amount due to Contractor = £64 178·63

## Appendix 1. Sketch of the layout of the Works

N

Littlesville — Bigsville

South Road

*PLAN OF ROAD PRIOR TO CONTRACT EXECUTION*

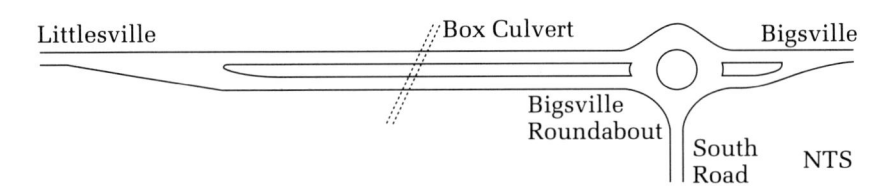

Littlesville — Box Culvert — Bigsville

Bigsville Roundabout

South Road   NTS

*PLAN OF ROAD AFTER CONSTRUCTION*

# Appendix 2. Original Clause 14(1) programme

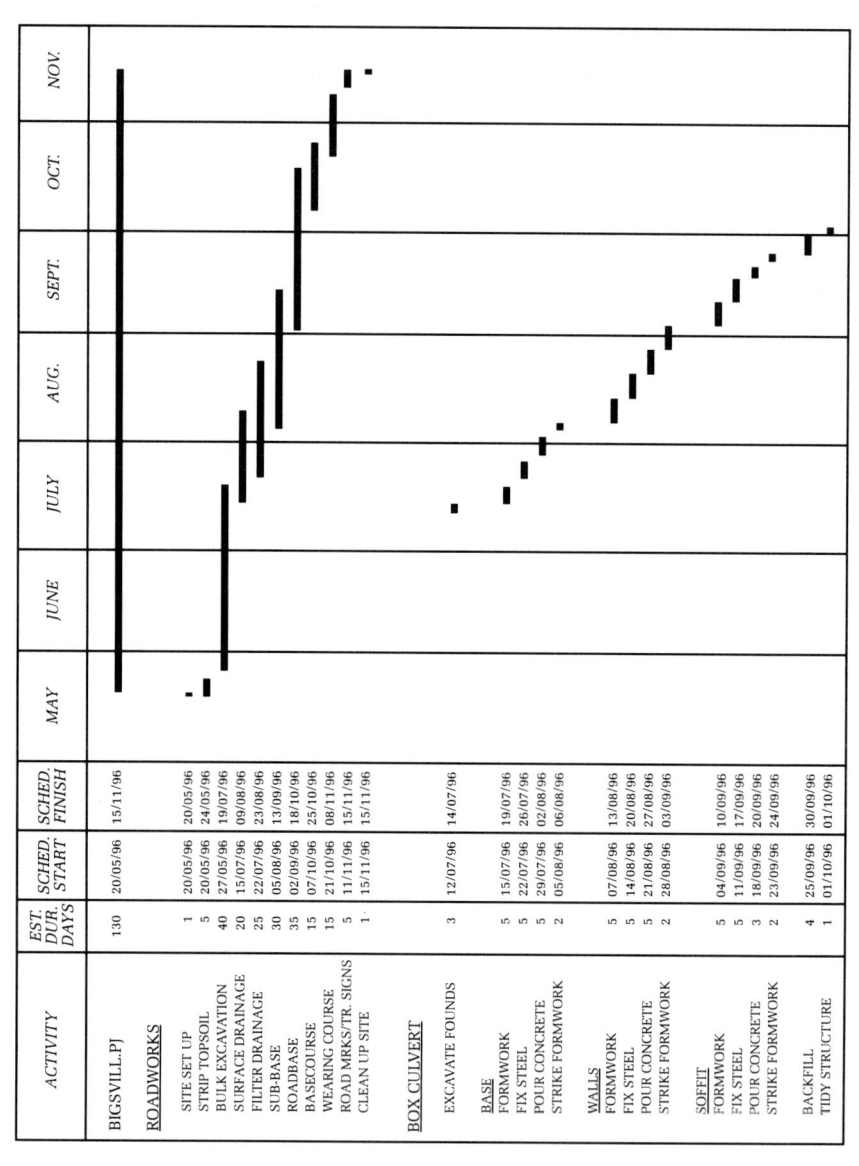

| ACTIVITY | EST. DUR. DAYS | SCHED. START | SCHED. FINISH |
|---|---|---|---|
| BIGSVILL.PJ | 130 | 20/05/96 | 15/11/96 |
| **ROADWORKS** | | | |
| SITE SET UP | 1 | 20/05/96 | 20/05/96 |
| STRIP TOPSOIL | 5 | 20/05/96 | 24/05/96 |
| BULK EXCAVATION | 40 | 27/05/96 | 19/07/96 |
| SURFACE DRAINAGE | 20 | 15/07/96 | 09/08/96 |
| FILTER DRAINAGE | 25 | 22/07/96 | 23/08/96 |
| SUB-BASE | 30 | 05/08/96 | 13/09/96 |
| ROADBASE | 35 | 02/09/96 | 18/10/96 |
| BASECOURSE | 15 | 07/10/96 | 25/10/96 |
| WEARING COURSE | 15 | 21/10/96 | 08/11/96 |
| ROAD MRKS/TR. SIGNS | 5 | 11/11/96 | 15/11/96 |
| CLEAN UP SITE | 1 | 15/11/96 | 15/11/96 |
| **BOX CULVERT** | | | |
| EXCAVATE FOUNDS | 3 | 12/07/96 | 14/07/96 |
| **BASE** | | | |
| FORMWORK | 5 | 15/07/96 | 19/07/96 |
| FIX STEEL | 5 | 22/07/96 | 26/07/96 |
| POUR CONCRETE | 5 | 29/07/96 | 02/08/96 |
| STRIKE FORMWORK | 2 | 05/08/96 | 06/08/96 |
| **WALLS** | | | |
| FORMWORK | 5 | 07/08/96 | 13/08/96 |
| FIX STEEL | 5 | 14/08/96 | 20/08/96 |
| POUR CONCRETE | 5 | 21/08/96 | 27/08/96 |
| STRIKE FORMWORK | 2 | 28/08/96 | 03/09/96 |
| **SOFFIT** | | | |
| FORMWORK | 5 | 04/09/96 | 10/09/96 |
| FIX STEEL | 5 | 11/09/96 | 17/09/96 |
| POUR CONCRETE | 3 | 18/09/96 | 20/09/96 |
| STRIKE FORMWORK | 2 | 23/09/96 | 24/09/96 |
| BACKFILL | 4 | 25/09/96 | 30/09/96 |
| TIDY STRUCTURE | 1 | 01/10/96 | 01/10/96 |

Note. This programme is based on working Mondays to Fridays only

## Appendix 3. List of drawings

(1) RD123/7/1A—Location plan and General layout
(2) RD123/7/2C—Details of utility equipment location
(3) RD123/7/3B—Carriageway cross sections—1
(4) RD123/7/4D—Carriageway cross sections—2
(5) RD123/7/5B—Carriageway longitudinal sections
(6) RD123/7/6B—Drainage details
(7) RD123/7/7C—Manholes and catchpits
(8) RD123/7/8B—Traffic signs and road markings
(9) RD123/7/10D—Box culvert—general layout
(10) RD123/7/11E—Box culvert—reinforcement details—1
(11) RD123/7/12E—Box culvert—reinforcement details—2

## Appendix 4. Correspondence related to claim

Letter C1

Unlimited Contracting Ltd<br>
28 The Yard<br>
Newtown Industrial Estate<br>
NEWTOWN NT9 9ZZ

12 July 1996

Mr R Sharpe
Sharpe & Sharpe
2 The Mews
NEWTOWN NT2 1BB

Dear Sir

**A1069—Bigsville to Littlesville**
**Construction of dual carriageway**
**Foundations to the box culvert**
**Notice of encountering adverse physical condition**

We have to advise you that we have encountered adverse physical conditions in the form of unacceptable material below the foundations to the box culvert. In our opinion this could not reasonably have been foreseen by an experienced contractor bearing in mind our obligations under Clause 11(1) and the information supplied in relation thereto. The effect of these adverse physical conditions is that we shall incur additional cost which would not have been incurred if these conditions had not been encountered.

We are excavating the unacceptable material and for that purpose have hired in a suitable excavator, lorries to take the excavated material to a tip some 8 km distance from the site, pumping equipment and an appropriate vibratory roller. We intend to excavate some 2 m below the underside of the foundation and propose to fill the void with rock fill some 200 mm to 0 mm down to a level 300 mm below the proposed formation of the structure. This material will be compacted in accordance with the requirements set out in Clause 612 of Volume 1: Specification for Highway Works. The remaining 300 mm will consist of Type 1 sub-base complying with

Clause 803 of Volume 1: Specification for Highway Works. We trust that our proposals meet with your approval.

This notice is issued pursuant to Clause 12(1) and Clause 52(4)(b) of the Conditions of Contract.

Yours faithfully


G Patton
Agent
for Unlimited Contracting Ltd

*Letter C2*

Sharpe & Sharpe
2 The Mews
NEWTOWN NT2 1BB

15 July 1996

Unlimited Contracting Ltd
28 The Yard
Newtown Industrial Estate
NEWTOWN NT9 9ZZ

Dear Sirs

**A1069—Bigsville to Littlesville**
**Construction of dual carriageway**
**Foundations to the box culvert**
**Requirement to provide estimate for measures**

I refer to your letter dated 12 July related to the above and note that you consider that encountering unacceptable material below the above constitutes adverse physical conditions which could not reasonably have been foreseen by an experienced contractor. Without prejudice to my powers under Clause 12(3), please provide me with an estimate of the costs thereof pursuant to Clause 12(2) of the Conditions of Contract.

Yours faithfully


R Sharpe
Engineer

*Letter C3*

                                                Unlimited Contracting Ltd
                                                        28 The Yard
                                        Newtown Industrial Estate
                                              NEWTOWN NT9 9ZZ

17 July 1996

Mr R Sharpe
Sharpe & Sharpe
2 The Mews
NEWTOWN NT2 1BB

Dear Sir

**A1069—Bigsville to Littlesville**
**Construction of dual carriageway**
**Foundations to the box culvert**
**Estimate of costs for dealing with adverse physical condition**

Further to your letter dated 15 July and our letter dated 16 July, please find attached an estimate of the costs of the measures which we are taking in order to deal with the adverse physical conditions at the above.

This estimate is provided pursuant to Clause 12(2) of the Conditions of Contract.

Yours faithfully


G Patton
Agent
for Unlimited Contracting Ltd

Enc

*Letter C4*

Sharpe & Sharpe
2 The Mews
NEWTOWN NT2 1BB

17 July 1996

Unlimited Contracting Ltd
28 The Yard
Newtown Industrial Estate
NEWTOWN NT9 9ZZ

Dear Sirs

**A1069—Bigsville to Littlesville**
**Construction of dual carriageway**
**Foundations to the box culvert**
**Confirmation of verbal instruction**

I confirm my verbal instruction to extend the depth of excavation of the unacceptable material to 3 m below the underside of the foundation level at the above. This instruction is issued pursuant to Clause 12(2) of the Conditions of Contract.

Yours faithfully


U Grant
Engineer's Representative

*Letter C5*

Unlimited Contracting Ltd
28 The Yard
Newtown Industrial Estate
NEWTOWN NT9 9ZZ

7 August 1996

Mr R Sharpe
Sharpe & Sharpe
2 The Mews
NEWTOWN NT2 1BB

Dear Sir

**A1069—Bigsville to Littlesville**
**Construction of dual carriageway**
**Foundations to the box culvert**
**Notice of claim due to adverse physical condition not reasonably foreseeable by an experienced contractor**

In our letter to you dated 12 July we advised you that the adverse physical condition in the form of unacceptable material below the foundations to the box culvert has resulted in us incurring additional cost which would not have been incurred if this condition had not been encountered.

Furthermore this event has delayed the execution of the Works. Our programme submitted to you on 14 May and subsequently approved by you on 21 May showed the erection of formwork to the base of the box culvert commencing on 15 July. This operation did not commence until 6 August and consequently the amount of the delay is 22 days.

We hereby request that pursuant to Clause 12(3) of the Conditions of Contract you grant an extension of time under the terms of Clause 44(1) of the Conditions of Contract and certify payment of this extra cost under the terms of Clause 60 of the Conditions of Contract.

We confirm that appropriate contemporary records have been kept to support this application for an extension of time and additional payment. Please note that this letter constitutes a notice required pursuant to Clause 52(4)(b) of the Conditions of Contract.

Yours faithfully


G Patton
Agent
for Unlimited Contracting Ltd

*Letter C6*

Sharpe & Sharpe
2 The Mews
NEWTOWN NT2 1BB

8 August 1996

Unlimited Contracting Ltd
28 The Yard
Newtown Industrial Estate
NEWTOWN NT9 9ZZ

Dear Sirs

**A1069—Bigsville To Littlesville**
**Construction of dual carriageway**
**Foundations to the box culvert**
**Conditions reasonably foreseeable**

I refer to your letter dated 17 July related to dealing with the unacceptable material below the box culvert to constitute adverse physical conditions which could not reasonably have been foreseen by an experienced contractor. I have to advise you that the nature of this material is entirely consistent with that suggested by the borehole logs for this area which were supplied to you at the time of tender. Accordingly I do not accept your view that this material constitutes adverse physical conditions since, on the basis of the aforementioned borehole logs and pursuant to your obligations under Clause 11 of the Conditions of Contract, they were reasonably foreseeable by an experienced contractor. I therefore propose to make no allowance for an extension of time and to evaluate the work at billed rates.

This decision is issued pursuant to Clause 12(4) of the Conditions of Contract.

Yours faithfully


R Sharpe
Engineer

*Letter C7*

Unlimited Contracting Ltd
28 The Yard
Newtown Industrial Estate
NEWTOWN NT9 9ZZ

16 August 1996

Mr R Sharpe
Sharpe & Sharpe
2 The Mews
NEWTOWN NT2 1BB

Dear Sir

**A1069—Bigsville to Littlesville**
**Construction of dual carriageway**
**Foundations to the box culvert**
**Notice of claim due to adverse physical condition not reasonably foreseeable by an experienced contractor**

We refer to your letter dated 8 August advising us that you consider that the unacceptable material encountered below the foundations to the box culvert was reasonably foreseeable by an experienced contractor. Item Number 600.3 relates to the excavation of acceptable material in the foundation of the box culvert. The Structures Earthworks Schedule indicates that the entire volume of excavation of the foundation with the exception of 10 m$^3$ consists of acceptable material. These quantities are reflected in Item Numbers 600.3 and 600.4 which cover the excavation of acceptable and unacceptable material respectively in the foundation of the box culvert. The volume of 10 m$^3$ was obviously included in the Bill of Quantities for the purposes of obtaining a rate without any evidence to substantiate the existence of unacceptable material. If it was not possible to deduce the existence of unacceptable material at the time when the Bill of Quantities was compiled then it cannot be plausible to conclude that its existence was reasonably foreseeable by an experienced contractor.

Accordingly, we request that you reconsider the view expressed in your letter and deem the event to constitute an adverse physical condition, order a Variation pursuant to Clause 51 of the Conditions of Contract and certify an extension of time pursuant to Clause 44(2) of the Conditions of Contract and extra costs pursuant to Clause 12(1) of the Conditions of Contract.

Yours faithfully


G Patton
Agent
for Unlimited Contracting Ltd

*Letter C8*

Sharpe & Sharpe
2 The Mews
NEWTOWN NT2 1BB

27 August 1996

Unlimited Contracting Ltd
28 The Yard
Newtown Industrial Estate
NEWTOWN NT9 9ZZ

Dear Sirs

**A1069—Bigsville to Littlesville**
**Construction of dual carriageway**
**Foundations to the box culvert**
**Conditions reasonably foreseeable**

I refer to your letter dated 16 August related to whether the unacceptable material below the foundation to the box culvert constitutes an adverse physical condition which could not reasonably have been foreseen by an experienced contractor. The fact is that there is an acknowledgement in the Contract in the form of an item in the Bill of Quantities that there is a possibility of unacceptable material. Furthermore, the boreholes indicate the clear prospect of encountering such material.

Your rates for excavation of acceptable material in structural foundations, unacceptable material in structural foundations for a depth of 0 to 3 m and also for excavation of soft spots below structural foundations are all the same.

The total quantity of soft spots included the Bill of Quantities has not been reached.

As a result of all the points made above I have to advise you that I do not consider that you are entitled to an extension of time thereto nor do I believe that this event constitutes an adverse physical condition. Consequently there is no cause to order a Variation pursuant to Clause 51 of the Conditions of Contract.

Yours faithfully


R Sharpe
Engineer

*Letter C9*

Unlimited Contracting Ltd
28 The Yard
Newtown Industrial Estate
NEWTOWN NT9 9ZZ

17 September 1996

Mr R Sharpe
Sharpe & Sharpe
2 The Mews
NEWTOWN NT2 1BB

Dear Sir

**A1069—Bigsville to Littlesville**
**Construction of dual carriageway**
**Foundations to the box culvert**
**Notice of claim due to adverse physical condition not reasonably foreseeable by an experienced contractor**

We refer to your letter dated 27 August advising us that you remain of the view that the unacceptable material encountered below the foundations to the box culvert was reasonably foreseeable by an experienced contractor. We have to advise you that we continue to hold the view expressed in previous correspondence. We are preparing details of our claim which will be submitted to you in due course.

Yours faithfully


G Patton
Agent
for Unlimited Contracting Ltd

## Appendix 5. Programme showing actual progress

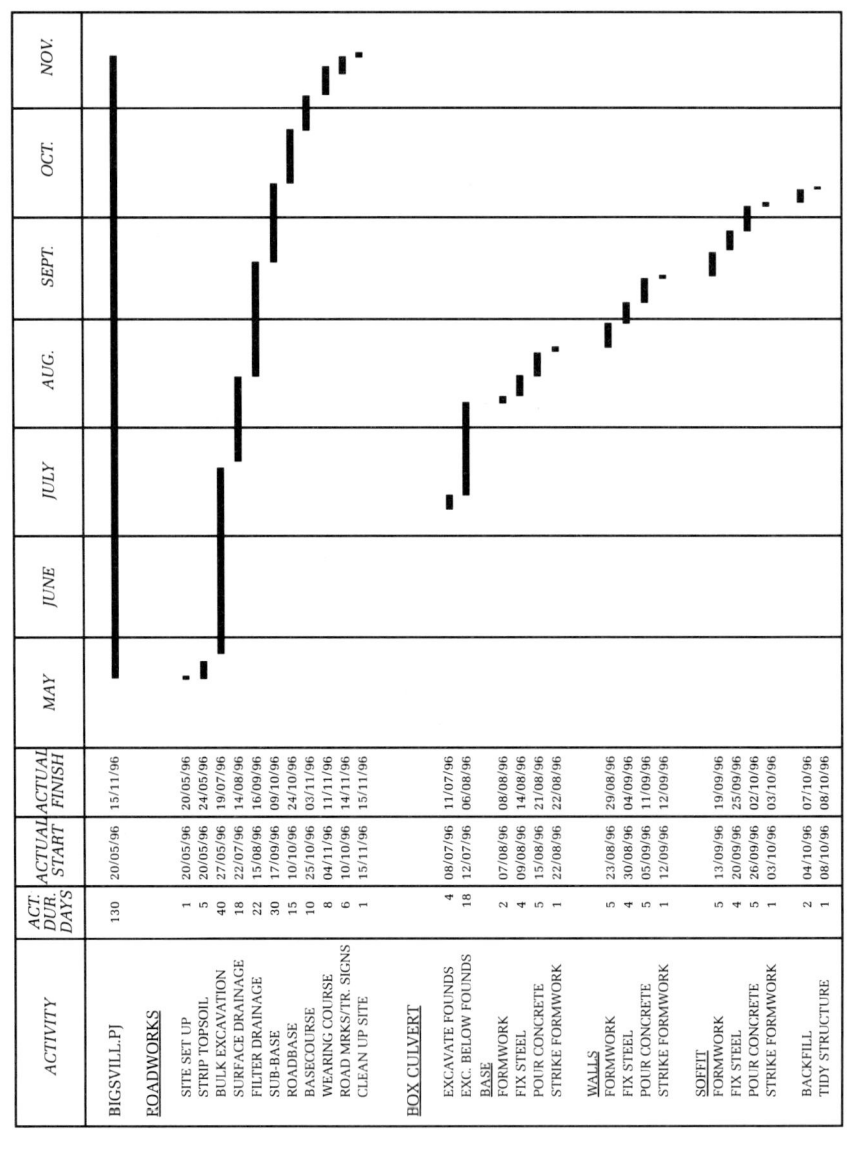

| ACTIVITY | ACT. DUR. DAYS | ACTUAL START | ACTUAL FINISH |
|---|---|---|---|
| BIGSVILL.PJ | 130 | 20/05/96 | 15/11/96 |
| **ROADWORKS** | | | |
| SITE SET UP | 1 | 20/05/96 | 20/05/96 |
| STRIP TOPSOIL | 5 | 20/05/96 | 24/05/96 |
| BULK EXCAVATION | 40 | 27/05/96 | 19/07/96 |
| SURFACE DRAINAGE | 18 | 22/07/96 | 14/08/96 |
| FILTER DRAINAGE | 22 | 15/08/96 | 16/09/96 |
| SUB-BASE | 30 | 17/09/96 | 09/10/96 |
| ROADBASE | 15 | 10/10/96 | 24/10/96 |
| BASECOURSE | 10 | 25/10/96 | 03/11/96 |
| WEARING COURSE | 8 | 04/11/96 | 11/11/96 |
| ROAD MRKS/TR. SIGNS | 6 | 10/10/96 | 14/11/96 |
| CLEAN UP SITE | 1 | 15/11/96 | 15/11/96 |
| **BOX CULVERT** | | | |
| EXCAVATE FOUNDS | 4 | 08/07/96 | 11/07/96 |
| EXC. BELOW FOUNDS | 18 | 12/07/96 | 06/08/96 |
| **BASE** | | | |
| FORMWORK | 2 | 07/08/96 | 08/08/96 |
| FIX STEEL | 4 | 09/08/96 | 14/08/96 |
| POUR CONCRETE | 5 | 15/08/96 | 21/08/96 |
| STRIKE FORMWORK | 1 | 22/08/96 | 22/08/96 |
| **WALLS** | | | |
| FORMWORK | 5 | 23/08/96 | 29/08/96 |
| FIX STEEL | 4 | 30/08/96 | 04/09/96 |
| POUR CONCRETE | 5 | 05/09/96 | 11/09/96 |
| STRIKE FORMWORK | 1 | 12/09/96 | 12/09/96 |
| **SOFFIT** | | | |
| FORMWORK | 5 | 13/09/96 | 19/09/96 |
| FIX STEEL | 4 | 20/09/96 | 25/09/96 |
| POUR CONCRETE | 5 | 26/09/96 | 02/10/96 |
| STRIKE FORMWORK | 1 | 03/10/96 | 03/10/96 |
| BACKFILL | 2 | 04/10/96 | 07/10/96 |
| TIDY STRUCTURE | 1 | 08/10/96 | 08/10/96 |

Note. Some of the activities shown above were carried out under 7 day working—see text for details

# Appendix 6. Dayworks sheets

**UNLIMITED CONTRACTING LTD**
Contract - A1069 Bigsville to Littlesville
Daywork Record Number - 51156

Work Description - Excavate and fill under foundation to box culvert

*1. Labour*

| Operative | Trade | Hours S | M | T | W | T | F | S | NPO | Total | Rate | Wages | Bonus | Expenses |
|---|---|---|---|---|---|---|---|---|---|---|---|---|---|---|
| P Graves | Foreman | | 9 | 9 | 9 | 9 | 9 | | 2.5 | 47.5 | 8.20 | 389.50 | 50.00 | 20.00 |
| S Hill | Ganger | | 9 | 9 | 9 | 9 | 9 | | 2.5 | 47.5 | 7.35 | 349.13 | 40.00 | 20.00 |
| M Landau | M/c Operator | | 9 | 9 | 9 | 9 | 9 | | 2.5 | 47.5 | 6.50 | 308.75 | 30.00 | 20.00 |
| G Morris | M/c Operator | | 9 | 9 | 9 | 9 | 9 | | 2.5 | 47.5 | 6.50 | 308.75 | 30.00 | 20.00 |
| B Bain | Banksman | | 9 | 9 | 9 | 9 | 9 | | 2.5 | 47.5 | 5.20 | 247.00 | 30.00 | 20.00 |
| P Lupus | Labourer | | 9 | 9 | 9 | 9 | 9 | | 2.5 | 47.5 | 5.20 | 247.00 | 30.00 | 20.00 |
| | | | | | | | | | Sub-totals | | | 1,850.13 | 210.00 | 120.00 |
| | | | | | | | | | | | | Total | | 2,180.13 |

## 2. Plant

| Item | Fleet No./ Registration | Time (hours/days) S | M | T | W | T | F | S | Total hours | Total days | Nominal size or capacity | Rate | Total |
|---|---|---|---|---|---|---|---|---|---|---|---|---|---|
| JCB 820 Excavator | E5 | | 9 | 9 | 9 | 9 | 9 | | 45 | | 20 t | 29.62 | 1,332.90 |
| Volvo A20 Dump Truck | D4 | | 9 | 9 | 9 | 9 | 9 | | 45 | | 20 t | 37.93 | 1,706.85 |
| Hamm DV3 Compactor | C1 | | 9 | 9 | 9 | 9 | 9 | | 45 | | 3.1 t | 13.86 | 623.70 |
| 76 mm Submersible Pump | N/A | 1 | 1 | 1 | 1 | 1 | 1 | 1 | | 7 | 76 mm | 5.55 | 38.85 |
| | | 24 | 24 | 24 | 24 | 24 | 24 | 24 | 168 | | 76 mm | 0.57 | 95.76 |
| 76 mm Suction Hose - 10 m | N/A | 1 | 1 | 1 | 1 | 1 | 1 | 1 | | 7 | 76 mm | 0.25 | 17.50 |
| 76 mm Delivery Hose - 5 m | N/A | 1 | 1 | 1 | 1 | 1 | 1 | 1 | | 7 | 76 mm | 0.22 | 7.70 |
| | | | | | | | | | | | | Total | 3,823.26 |

### 3. Material

| Material | Quantity | Rate | Total |
|---|---|---|---|
| Rock Fill - 300 mm to Dust | 586.32 t | 8.50 | 4,983.72 |
| Whin Dust | 45.11 t | 7.40 | 333.00 |
|  |  |  |  |
|  |  |  |  |
|  |  |  |  |
|  |  |  |  |
|  |  | Total | 5,316.72 |

*Certified as a true record*

Signed for Unlimited Contracting Ltd:-

*G S Patton*

Signed for Engineer:-

*W J Grant*

**FOR RECORD PURPOSES ONLY**

Date ___29 July 1996___

*Total Labour + Plant + Materials = £11,320.11*

Note. This sheet is for one week only. Similar sheets would be presented for the other periods involved which takes the total to, say, £26 279·44.

## Appendix 7. Non-productive overtime costs

*Surfacing sub-contractor's costs*

**UNLIMITED CONTRACTING LTD**
Contract - A1069 Bigsville to Littlesville
Sub-Contractor's Daywork Record Number SC1054 for Billiard Table Surfacing Ltd
Work Description - Various Surfacing Operations
Week Ending:- 19 October 1996

### 1. Labour

| Operative | Trade | S | M | T | W | T | F | S | NPO Total | Rate | Wages | Bonus | Expenses |
|---|---|---|---|---|---|---|---|---|---|---|---|---|---|
| J Phelps | Supervisor | 5 | 1 | 1 | 1 | 1 | 1 | | 10 | 10.25 | 102.50 | | |
| D Briggs | Foreman | 5 | 1 | 1 | 1 | 1 | 1 | | 10 | 7.50 | 75.00 | | |
| R Hand | Screwman | 5 | 1 | 1 | 1 | 1 | 1 | | 10 | 7.00 | 70.00 | | |
| B Collier | Paver Op. | 5 | 1 | 1 | 1 | 1 | 1 | | 10 | 7.00 | 70.00 | | |
| C Carter | Roller Driver | 6 | 2 | 2 | 2 | 2 | 2 | | 16 | 6.50 | 104.00 | | |
| W Armitage | Roller Driver | 5 | 1 | 1 | 1 | 1 | 1 | | 10 | 6.50 | 65.00 | | |
| D Roberts | Raker | 5 | 1 | 1 | 1 | 1 | 1 | | 10 | 6.10 | 61.00 | | |
| D Lambert | Raker | 5 | 1 | 1 | 1 | 1 | 1 | | 10 | 6.10 | 61.00 | | |
| M Davis | Labourer | 5 | 1 | 1 | 1 | 1 | 1 | | 10 | 6.10 | 61.00 | | |
| | | | | | | | | | | Sub-totals | 669.50 | | |
| | | | | | | | | | | Add 88% | 589.16 | | |
| | | | | | | | | | | | | Total | 1,258.66 |

*Note. This sheet is for one week only. Similar sheets would be presented for the other periods involved which takes the total to, say, £6,551·53.*

### Contractor's own costs

*Note. This would be similar to that shown above and amount to, say, £5,456·44.*

# Index of Standard Letters

| Letter No. | Clause No. | Description | Page No. |
|---|---|---|---|
| 1 | 7(2) | Request for further drawings and/or instructions | 21 |
| 2 | 7(3) | Notice of delay and extra cost as a result of a delay in the issue of drawings/instructions | 22 |
| 3 | 12(1) | Notice of encountering adverse physical condition/artificial obstruction and measures proposed/being taken | 30 |
| 4 | 12(2) | Provision of ordered estimate of costs of actual or proposed measures to deal with adverse physical conditions/artificial obstructions | 31 |
| 5 | 12(3) | Notice of claim as a result of encountering adverse physical condition/artificial obstruction | 32 |
| 6 | 5/13(1) | Notification of claim as a result of delay and/or disruption resulting from an Engineer's instruction | 36 |
| 7 | 14(6) | Notice of claim as a result of an unreasonable delay in approval of proposed order of procedure | 41 |
| 8 | 14(6) | Notice of claim due to limitations imposed by the provision of design criteria not reasonably foreseeable by an experienced Contractor at the time of tender | 42 |
| 9 | 17(2) | Notice of claim as a result of an error in the setting out data supplied by the Engineer or Engineer's Representative | 44 |

| Letter No. | Clause No. | Description | Page No. |
|---|---|---|---|
| 10 | 20(2) | Notice of claim as a result of damage loss or injury occurring under an Excepted Risk | 47 |
| 11 | 27(6) | Notice of claim as a result of compliance with a variation related to utility operations | 52 |
| 12 | 31(2) | Notice of claim as a result of delay and extra cost resulting from the operations of another contractor, authorised authority or statutory body | 54 |
| 13 | 36(2)&(3) | Notice of claim as a result of amount or types of sampling and testing rising above that suggested in the tender documents | 57 |
| 14 | 38(2) | Notice of claim as a result of damage loss or injury occurring under an excepted risk | 59 |
| 15 | 40(1) | Notice of claim as a result of suspension of the works or part thereof | 62 |
| 16 | 42(1) | Notice of claim as a result of non-possession of all or part of the site | 64 |
| 17 | 42(1) | Notice of claim as a result of late possession of part of the site due to operations by another contractor | 65 |
| 18 | 44(1)&(2) | Application for extension of time | 71 |
| 19 | 48(1) | Application for Certificate of Completion for the whole of the Works | 78 |
| 20 | 48(2)(a) | Application for Certificate of Completion for a section of the Works | 79 |
| 21 | 48(2)(b) | Application for Certificate of Completion for a substantial part of the Works | 80 |
| 22 | 50 | Contractor's request for payment for searches tests or trials | 81 |
| 23 | 51(1)&(2) | Confirmation of Variation ordered orally by the Engineer | 86 |
| 24 | 52(2) | Request for the Engineer to vary a rate/price for ordered Variation | 93 |
| 25 | 52(4)(d) | Notice of intention to claim for a higher rate/price than that notified | 94 |

# INDEX OF STANDARD LETTERS | 223

| Letter No. | Clause No. | Description | Page No. |
|---|---|---|---|
| 26 | 52(4)(d) | First interim intimation of value of claim | 95 |
| 27 | 55(2) | Notice of claim for increased rate and extension of time in relation to errors/omissions from bill of quantities | 97 |
| 28 | 56(2) | Request for re-rating as a result of a change in quantities | 99 |
| 29 | 56(2) | Notice of claim as a result of a notification of rate related to a change in quantities | 100 |
| 30 | 59A(2)(b) | Notice of recovery of charges and profit related to the cancellation of a prime cost item | 105 |
| 31 | 59A(3)(b) | Notice of claim as a result of a Nominated Sub-Contractor refusing to accept specified provisions | 106 |
| 32 | 59B(4)(b) | Notice of claim for extension of time and recovery of additional cost incurred as a result of the forfeiture of a contract with a Nominated Sub-Contractor | 111 |
| 33 | 60(1) | Submission of monthly statement | 119 |
| 34 | 60(3) | Submission of final account | 120 |
| 35 | 61(1) | Request for Engineer to issue Maintenance Certificate | 121 |

# General Index

Abrahamson   28, 69, 125
Acceleration   69
Adjudication   133
Adverse physical conditions, see
    chapter 5   16, 25–33, 132,
    148, 188
Adverse physical conditions and
    artificial obstructions   25
Adverse weather conditions   28, 61,
    68, 70, 75
Amendments to the *Manual of
    Contract Documents for Highway
    Works*   143
Amendments to the *ICE Conditions of
    Contract*   138
Amount of wages   148
Annex A   165
Anti-collusion certificate   147
Appendices
    Lettered   157–159, 162
    Numbered   159, 161, 163, 164
Appendix 0/1   163, 164, 165
Appendix 0/2   163, 164, 165
Appendix 0/3   164, 177
Appendix 0/4   164
Appendix 0/5   164, 166, 177
Appendix 1/5   56, 166
Appendix 1/6   167
Appendix 1/14   160
Appendix 1/16   174
Appendix to the Form of Tender   9,
    66, 77, 116, 146, 148
Application for extension of time   71
Application for Certificate of
    Completion   78, 79, 80
Arbitration   122–127, 148–149, 184,
    186
    Act   127
    Engineer as witness   124
    Engineer's decision, time for   123

ICE Procedure (1983)   124, 127
President or Vice-President to
    act   123
Time for Engineer's decision   123
Arbitrator   75, 125
    award by   127
Artificial obstructions   16, 25–33,
    132, 148
Assessment at due date for
    completion   68
Attending for measurement   98
Avoiding claims   4

Base lending rate   118, 148
Bill of quantities   85
    correction of errors   96
    for Highway Works   156, 160, 166,
        167–175, 177
    omissions from the   96
*Bills of Quantities for Highway
    Works*   156, 160, 166, 167–175,
    177
    Preparation of   156, 170
    Units and Methods of
        Measurement   171
Billed quantities   85
Billed rates   90
    decrease in   90, 98
    increase in   90, 98
Breach of a common law duty in
    tort   6
Breach of sub-contract (nominated
    sub-contract)   104

Care of the Works   45
Certificate
    anti-collusion   147
    final   117, 125
    maintenance   116, 117, 119,
        120, 121

# GENERAL INDEX | 225

prompt payment 147
Certificate of Completion  45, 46–47, 70, 74, 76–80, 92, 118, 148
   Application for  78, 79, 80
Certificate of Completion of Works  76–80
Certificates and payment  92
Changes in quantities  68, 83
Changes to the *5th Edition of the ICE Conditions of Contract*  138, 145
Claims
   associated with the *Manual of Contract Documents for Highway Works*  175
   avoiding  4
   contractual  8
   emanating from the Bill of Quantities  177
   emanating from the Method of Measurement  181
   emanating from the Specification  177
   estimated amount of  115
   *ex gratia*  7
   for extensions of time  133
   interim intimation of value of  95
   nature of  8–9
   notice of  26, 41–45, 47, 52, 54, 57, 62, 64, 65, 88, 93, 96, 97, 100, 111, 132
   philosophy of  1–4
Clause 1  139
Clause 1(3)  17
Clause 1(5)  132, 170
Clause 2  139
Clause 2(2)  34
Clause 5  17–18, 36
Clause 7  19–23, 147
Clause 7(1)  67
Clause 7(3)  67
Clause 7(4)  19
Clause 11  23–25, 192, 211
Clause 11(1)  204
Clause 12  25–33, 35, 132, 133, 148, 192
Clause 12(1)  30, 132, 204
Clause 12(2)  31, 206, 207, 208
Clause 12(3)  32, 206, 209
Clause 12(4)  211
Clause 13  33–37
Clause 13(1)  85
Clause 13(3)  18, 40, 85
Clause 14  37–43, 64, 89, 154
Clause 14(1)  40, 197, 202
Clause 17  43–45
Clause 17(2)  44
Clause 20  45–48
Clause 20(3)  61
Clause 27  17, 48–52, 155
Clause 27(1)  147
Clause 31  53–55
Clause 36  69
Clause 38  58–60
Clause 39  59
Clause 40  27, 60–62
Clause 41  63, 64
Clause 42  63–66, 147, 155
Clause 43  66–67
Clause 44  20, 27, 40, 51, 64, 67–75
Clause 44(1)  44, 45, 66, 133, 209
Clause 44(2)  213
Clause 44(3)  75
Clause 44(4)  75
Clause 47  63, 70, 71–75
Clause 48  47, 74, 76, 80, 121
Clause 48(2)  74
Clause 49  81, 121
Clause 49(1)  77
Clause 50  81–82, 121
Clause 51  20, 27, 34, 35, 41, 61, 82–86, 96, 110, 212, 214
Clause 51(1)  68, 196
Clause 51(3)  68
Clause 52  27, 51, 86–96, 98
Clause 52(3)  148, 155, 196
Clause 52(4)  20, 23, 26, 40, 64, 132
Clause 52(4)(a)  14, 27, 35, 93, 99
Clause 52(4)(b)  14, 27, 35, 45, 66, 93, 205, 210
Clause 55  96–98
Clause 56  98–101, 155
Clause 56(2)  92, 100
Clause 57  148
Clause 59A  101–107
Clause 59A(2)  110
Clause 59B  107–111
Clause 60  20, 23, 27, 51, 66, 75, 92, 112–120, 139, 148, 209
Clause 60(2)  64, 148
Clause 60(4)  74, 77, 155
Clause 60(6)  116, 118, 148, 199
Clause 61  120–122
Clause 63  110
Clause 63(1)  148
Clause 66  118, 122, 127, 139, 148, 155

Clause 69   139
Clause 72   139
Clause 74   149
Clause 76   149, 154
Clause 77   149, 155
Clause 78   149, 155
Clause 79   149, 155
Clause 80   149, 155
Clause 81   155
Clause 82   155
Clause 83   155
*Clydebank Engineering and Shipbuilding Co v Don Jose Ramos Yzquierdo-y-Castaneda (1905) AC6*   8, 74
Commencement of Works   63
Common law breach of contract   7
Completion, Certificate of   45, 46–47, 70, 74, 76–80, 92, 118, 148
  Application for   78, 79, 80
Completion, Date for   74
Completion of other parts of works   76
Completion of Sections   46, 74, 146
Completion of sections and occupied parts   76
Conditions reasonably foreseeable   26
Condition precedent   16
*Conditions of Contract*, see Chapter 2   14, 147, 170
  publication history of the ICE   10
*Constructing the Team*, the Latham Report   133
Contingencies   151
Contra proferentem   5, 18
Contract specific information   159, 167
Contract specific requirements   161
Contractor responsible for nominated sub-contracts   103
Contractor to comply with other obligations of Act   50
Contractor to search   81
Contractual claims   8
Controlled land   50
Copy certificate for Contractor   115
Correction of errors   96
Correction and withholding of certificates   115
Cost   132–133
  head office   132, 194
  of preparing claim   200
  overhead   132, 170
    percentage addition to   132
Cost of execution of work of repair   81
Cost of samples   55
Cost of searches, tests or trials   81
Cost of tests   56

Damages   197
Damages against (nominated) sub-contractor   104
Damages not a penalty   73
Date for Commencement of the Works   63, 64, 69
Date for Completion   69, 74, 197
Days (legal meaning)   41
Dayworks   27, 87, 90, 133, 151, 153, 197, 217–219
  Schedule of   90, 148, 151–153, 155
Decrease of rate   91
Decreased quantities   83
Deduction of liquidated damages   73
Defects   81
Defects liability period   74
Delay and disruption   27, 28, 36, 39, 75, 133, 174, 176
Delay and extra cost   25, 33, 38, 53, 108
Delay due to
  other Contractor   54
  authorised authority   54
  statutory body   54
Delay in issue   19
Delays approval, Engineer   40
Delays attributable to variations   50
Delict   6
Design criteria   38
*Design Manual for Roads and Bridges*   144, 154
Direction by Engineer   102
Dispute resolution   124
DMRB, see *Design Manual for Roads and Bridges*
Documents mutually explanatory   17
Drainage and service ducts   174

Earthworks   174
Earthworks balance   174
Earthworks Schedule, Roadworks   174
Earthworks Schedule, Structures   174

Emden formula   196
Emergency works   50
End product specification   157
Engineer as witness   124
Engineer delays approval   40
Engineer to fix rates   87
Engineer's action upon objection (to nominated sub-contractor)   102
Engineer's action upon termination (of nominated sub-contract)   108
Engineer's consent   38
Engineer's decision   118, 124
  arbitration—time for   123
  effect on Contractor and Employer   122
Engineer's instruction   37
Error in setting out   44
Errors of description   96
Estimated amount of claims   115
*Ex gratia* claims   7
Examination of work before covering up   58
Excepted risks   46, 47, 61
Exceptional adverse weather conditions   28, 61, 68
Express term   7
Extension of time   8, 40, 44, 67–71, 75
  assessment at due date for completion   68
  final determination of extension   68
  interim assessment of extension   68

Facilities for other Contractors   53
Failure to commence street works   49
Fair valuation   89
Finance charges   199
Final determination of extension   68
Final account   113, 117, 119, 121
  submission of   120
Final certificate   117, 125
Final quantities   85
Forfeiture   110, 148
Forfeiture of sub-contract   107
Form of bond   147
Forms of agreement by deed   147
*Functus officio*   125
Furmston   28
Further drawings and instructions   19

General Directions   170, 171, 181
*Glenlion Construction v. Guinness Trust* (1988) 38 BLR 89   20, 67

Head office costs   132, 194
Highway Construction
  Details   156, 173
  Implementation of   156
Housing Grants, Construction and Regeneration Act   133
Hudson   6, 13, 28, 41, 85
Hudson formula   196

*ICE Conditions of Contract*, see Chapter 2   14, 147, 170
  amendments to the   138
  publication history of the   10
ICE Arbitration Procedure (1983)   124, 127
*Implementation of Highway Construction Details*   156
*Implementation of Specification for Highway Works and Notes for Guidance*   155
Implied term   7
Increase or decrease of rate   92, 98
Increased quantities   68, 83
Inspection of site   23
Instructions   37
  oral   84
Instructions for tendering   145, 146
Insurance   74, 77
Interest on overdue payments   114, 116, 118
Interim assessment of extension   68
Interim certificate of payment   148
Interim intimation of value of claim   95
Item Coverages   53, 150, 169, 170, 171, 181
Item Descriptions   169, 170, 171, 174, 181
Itemisation Tables   170, 172, 174
  Features in   170, 172
  Groups in   169, 171

Joyce   138, 170, 172

Lane rental contracts   154
Latham Report   133
Lettered Appendices   157–159, 162

Library of Standard Item Descriptions
    for Highway Works   167,
    174, 177
Liquidated damages   8, 63, 70, 71,
    74, 77, 146, 197
  damages not a penalty   73
  deduction of   73
  reimbursement of   73
Liquidated damages for sections   72
Liquidated damages for whole of
    Works   71
List A   164
List B   164


Maintenance certificate   116, 117,
    119, 120–121
Maintenance period   77, 125
Maker's rated nominal weight of
    machine   27
*Manual of Contract Documents
    for Highway Works*, see
    Chapter 3   10, 91, 140
  amendments to   143
  current contents of   141
  publication history of the   11
  schedule of pages and relevant
    publication dates   145, 150, 158,
    161, 165, 171
Marginal Headings   150, 169, 181
Materials   153
Materials on site   119
MCD, see *Manual of Contract
    Documents for Highway Works*
Measurement and valuation   98
Measures to be taken   25
*Merton London Borough v.
    Leach (Stanley Hugh)* (1985)
    32 BLR 51   7
Method of Measurement for Highway
    Works   5, 53, 140, 148, 167, 168,
    169–175, 181
  General Principles   169
  Preparation of Bill of
    Quantities   170
Method specification   157
Methods of construction   37
Minimum certificate value   116
Minimum lending rate   118, 148
Mode and manner of
    construction   33
Model Contract Document   15, 23,
    145–155

Model Contract Document for Highway Works—England   145–153
Model Contract Document for
    Highway Works—Northern
    Ireland   154–155
Model Contract Document for
    Highway Works—
    Scotland   153–155
Model Contract Document for
    Highway Works—Wales   154
Model Contract Document for Major
    Works and Implementation
    Requirements   143–156, 160,
    162, 166
Money and Hodgson   140
Monthly payments   64, 112
Monthly statements   27, 112,
    158, 160
  submission of   119


Nature of claims   8–9
New Roads and Street Works Act   50,
    147, 155
Nominated Sub-Contractors
  breach of sub-contract   104
  Contractor responsible for   103
  delay and extra cost   108
  direction by Engineer   102
  Engineer's action upon
    objection   102
  Engineer's action upon
    termination   108
  forfeiture of sub-contract   107
  objection to nomination   101
  payments   103, 115
  recovery of Employer's loss   110
  termination of sub-contract   107
  termination without consent   109
Non-productive overtime   152,
    197, 220
Non-standard items   5, 140, 150, 176,
    177, 181
Notes for Guidance on the Method of
    Measurement for Highway
    Works   167, 172–174, 175, 192
Notes for Guidance on the
    Specification for Highway
    Works   156, 160–167, 174, 177
Notice by contractor   19
Notice enforcing forfeiture   110
Notice of claim   26, 41–45, 47, 52,
    54, 57, 62, 64, 65, 88, 93, 96, 97,
    100, 111, 132

# GENERAL INDEX | 229

Notice of delay   54, 59, 62, 64, 65
Notice to concur   125
Notice under
  Clause 5   36
  Clause 7(2)   21
  Clause 7(3)   22
  Clause 12(1)   30
  Clause 12(2)   31
  Clause 12(3)   32
  Clause 13(1)   36
  Clause 14(6)   41, 42
  Clause 17(2),   44
  Clause 20(2)   47
  Clause 27(6)   52
  Clause 31(2)   54
  Clause 36(2)/(3)   57
  Clause 38(2)   59
  Clause 40(1)   62
  Clause 42(1)   64, 65
  Clause 44(1)/(2)   71
  Clause 48(1)   78
  Clause 48(2)(a)   79
  Clause 48(2)(b)   80
  Clause 50   81
  Clause 51(1)/(2)   86
  Clause 52(2)   93
  Clause 52(4)(d)   95
  Clause 55(2)   97
  Clause 56(2)   99, 100
  Clause 59A(2)(b)   105
  Clause 59A(3)(b)   106
  Clause 59B(4)(b)   111
  Clause 60(1)   119
  Clause 60(3)   120
  Clause 61(1)   121
Notices   16
Notices by Contractor to
  Employer   49
Notifications by Employer to
  Contractor   48
Numbered Appendices   159, 161,
  163, 164
NRSWA   50, 147, 155

Objection to nomination of sub-
  contractors   101
Obstructions, Artificial   16, 25–33,
  132, 148
Omissions from the bill of
  quantities   96
One copy of documents to be kept on
  site   19
Oral instructions   84

Ordered variations   20, 35, 68,
  82, 89
Ordered variations to be in
  writing   82
Other special circumstances of any
  kind whatsoever   45, 68, 69
Overdue payments, interest on   114
Overhead costs   132, 170

Payment
  copy certificate for Contractor   115
  correction and withholding of
    certificates   115
  final account   113, 117, 119, 121
  Interim certificate   148
  interest on overdue   114, 116, 118
  minimum certificate value   116
  monthly   64, 112
  monthly statements   27, 112,
    158, 160
  retention   74, 77, 113, 118, 148
  value of goods and materials   115
Payments for nominated sub-
  contracts   103, 115
Percentage addition to cost   132
Percentage adjustment   152
Period of maintenance   77, 125
Philosophy of claims   1–4
Physical conditions, adverse, see
  Chapter 5   16, 25–33, 132, 148
Plant   27, 153
Possession of site   147
Preambles to Bill of Quantities   148,
  150, 170–171, 181
Preamble to the Specification   149,
  161, 165
Preparation of Bill of Quantities for
  Highway Works   156, 170
President or Vice-President to act (to
  appoint arbitrator)   123
Price fluctuation clause   7, 147, 149
Prime cost item   104
Profit   105, 132, 194
Programme   37–43, 85, 198, 202, 216
Programme to be furnished   37
Prompt payment certificate   147
Prospectively maintainable
  highway   50
Provisional sum   104, 110
Public Utilities Street Works Act   48–
  52, 147, 155
Publication history of the *ICE*
  *Conditions of Contract*   10

Publication history of the *Manual of Contract Documents for Highway Works*  11

Quality of materials and workmanship and tests  55
Quantities  96, 170, 181
*Quantum meruit*  6

Recent legislation to provide for adjudication  133
Recovery of Employer's loss (in nominated sub-contract)  110
Reimbursement of liquidated damages  73
Reinstatement of ground  77
Request for re-rating  100
Request to vary rate  93
Responsibility for reinstatement  46
Responsibility unaffected by approval  39
Results specification  157
Retention  74, 77, 113, 116, 118, 148
Retention money, payment of  113
Revision of programme  37, 39
Roadworks Earthworks Schedule  174
Rogue items  5, 140, 150

Sale of Goods Act  7
Samples  56, 167
Schedule of Dayworks  90, 147, 151–153, 155
Schedule of Pages and Relevant Publication Dates  150, 158, 161, 165, 171
Searches tests or trials  81
Sectional completion  46, 74, 146
Service locations  174
Serving of notices by Employer  49
Setting out  43
Settlement of disputes—arbitration  122
Similar character  89
Similar conditions  89
Special requirements  149
Specifications
   end product  156
   method  156
   results  156
Specification for Highway Works  56, 140, 142, 155, 156, 157–160, 161, 163, 164, 165, 166, 173, 177, 204

Standard forms of contract  3, 8
Standard letter pursuant to
   Clause 5  36
   Clause 7(2)  21
   Clause 7(3)  22
   Clause 12(1)  30
   Clause 12(2)  31
   Clause 12(3)  32
   Clause 13(1)  36
   Clause 14(6)  41, 42
   Clause 17(2)  44
   Clause 20(2)  47
   Clause 27(6)  52
   Clause 31(2)  54
   Clause 36(2)/(3)  57
   Clause 38(2)  59
   Clause 40(1)  62
   Clause 42(1)  64, 65
   Clause 44(1)/(2)  71
   Clause 48(1)  78
   Clause 48(2)(a)  79
   Clause 48(2)(b)  80
   Clause 50  81
   Clause 51(1)/(2)  86
   Clause 52(2)  93
   Clause 52(4)(d)  95
   Clause 55(2)  97
   Clause 56(2)  99, 100
   Clause 59A(2)(b)  105
   Clause 59A(3)(b)  106
   Clause 59B(4)(b)  111
   Clause 60(1)  119
   Clause 60(3)  120
   Clause 61(1)  121
Statutory body  54
Structures Earthworks Schedule  174
Sub-Heading  169, 171
Subcontractor (nominated)  101–111
Substantial completion  74
Sufficiency of tender  24
Suffixes to Clauses  166
Summary of claims procedure  132–133
Summary of provisions of clauses  127–132
Suspension  27, 61–62
Suspension lasting more than three months  60
Suspension of work  60

Termination without consent (of nominated sub-contract)  109

Termination of (nominated) sub-
    contract   107
Tests   56
Testing, typical details   166–167
Time for Completion   9, 66–67
Tort   6–7

Unfulfilled obligations   121
*Ultra vires*   34
Uncovering and making
    openings   58
Undertaker   51, 174
Units and Methods of
    Measurement   171

Valuation   89
Valuation of ordered varia-
    tions   27, 86, 89–96
Value of goods and materials   115
Variation   20, 27, 35, 51, 68, 82–94,
    213, 214

changes in quantities   83
daywork   87
Engineer to fix rates   87
Variation (*cont.*)
    notice of claims   88
    ordered   82, 89
    ordered, to be in writing   82
    valuation of ordered   27, 86

Wayleaves, etc   64, 147
Weather conditions   28, 61, 68,
    70, 75
Work to be to satisfaction of
    Engineer   33
Working rule agreement   148

*Yorkshire Water Authority v. (Sir
    Alfred) McAlpine Ltd* (1985)
    32 BLR 114   41, 85